Dancing AT THE Dead Sea

A JOURNEY TO THE HEART OF ENVIRONMENTAL CRISIS

Alanna Mitchell

eden project books

TRANSWORLD PUBLISHERS
61–63 Uxbridge Road, London W5 5SA
a division of The Random House Group Ltd

RANDOM HOUSE AUSTRALIA (PTY) LTD
20 Alfred Street, Milsons Point, Sydney,
New South Wales 2061, Australia

RANDOM HOUSE NEW ZEALAND LTD
18 Poland Road, Glenfield, Auckland 10, New Zealand

RANDOM HOUSE SOUTH AFRICA (PTY) LTD
Endulini, 5a Jubilee Road, Parktown 2193, South Africa

Published 2005 by Eden Project Books
a division of Transworld Publishers

A catalogue record for this book is available
from the British Library.
ISBN 1903 919630

Typeset in 11.5/16.5 Fairfield by
Falcon Oast Graphic Art Ltd

Printed in Great Britain by
Bookmarque Ltd, Croydon, Surrey

1 3 5 7 9 10 8 6 4 2

Papers used by Eden Project Books are made from wood grown
in sustainable forests. The manufacturing processes conform to
the environmental regulations of the country of origin.

To my children, Calista and Nicholas

CONTENTS

INTRODUCTION Dancing at the Dead Sea 9

ONE The Devil's Chaplain 27

TWO In Search of Lemurs 65

THREE Reading the Secrets of the Fossils 99

FOUR Parched Oasis 121

FIVE Sunken Graves of the High Arctic 147

SIX Where the Rainforest Goes On For Ever 171

SEVEN Iceland's New Power 209

EIGHT Ocean's End 227

EPILOGUE Writing New Legends 263

ACKNOWLEDGEMENTS 275

SELECTED REFERENCES 281

INDEX 291

INTRODUCTION

Dancing AT THE Dead Sea

I remember the heat of that night more than anything else, when the sluggish winds of Africa met up with the fevered South Asian air. Heat so profound it made children sigh in their sleep and adults squirm for embrace. We had just arrived by bus from an international conference on conservation in Amman for a royal feast on the Jordanian bank of the Dead Sea, guests of Queen Noor, who is a patron of the World Conservation Union.

Our hosts had set up a tent on the banks of the Dead Sea and covered the sand with lush Persian carpets whose ancient stories seemed to leap to life under our feet. An orchestra played, the musicians buttoned down in black. Waiters milled through the crowd offering endless streams of jewel-coloured fresh juices. All along the perimeter of the tent, long tables groaned with food, from hamburgers to spiced lentils to towers of honey-soaked Middle Eastern sweets.

The Dead Sea beckoned in the distance, across the field of carpets. As if summoned, I floated through the crowds to the bank to gaze at it. Suddenly, I had an irresistible urge to feel its salt on my skin. Bright gunfire broke out on the Israeli side opposite, towards Hebron and Bethlehem, part of the bitter battle that defines the Middle East of our time.

Even that failed to shock me back to reality. My head was filled – then, as it is now that I am writing about it – with the stories of that place. With what it has meant over time to the people who have known it.

To biblical writers, the Dead Sea was a symbol of devastation, an emblem to the ancients of how powerfully God's wrath could crush man's transgressions.

The sea, which is the lowest point on the planet and is so laced with salt that no human can drown in its depths, was the setting of the story of the city of Sodom, according to modern scholars of the Bible's geography.

At the conclusion of the story Lot's wife – fatally disobedient – is turned into a pillar of salt. Even today, there are pillars of salt scattered in the cliffs around the Dead Sea, pushed up through the land by the intense pressure of air and water. Some take that as evidence that biblical writers were describing this precise bit of the planet.

Ancient heartbreak permeated the place. It seemed to mingle with the modern tragedies under way. The sea itself is dying a second death, its water levels falling far faster than nature can replenish them as humans harness the mighty Jordan River for crops and industry. And I was mired in despair anyway. I had spent the week before the feast in the deserts at the fabled oasis of Azraq, cataloguing some of the worst ecological destruction I was ever to find on the planet.

But just as hope seemed unwarranted, just as I thought I could not possibly take any more despair, I had shown up at an auditorium in Amman filled with scientists and politicians and been handed a prize for best environmental reporter in the world for a story I had written for Canada's national newspaper *The Globe and Mail*. Heady stuff.

One part of the prize was a mosaic, constructed in the ancient Jordanian manner of piecing together tiny, irregularly shaped bits of stone to create the potent image of a 'woman-tree', symbol of life. It was achingly real: I had spent part of the evening lugging it with me over the torrid night sands, struggling not to drop it and worrying about how to get it home on the plane.

The second part of the prize was a fellowship to study at Oxford University. It was this part that had me dreaming so big, so deep, that it scared me. All of a sudden, I could feel the whole world opening up before me. I shook off the gloom, able finally to taste the urgent adventures that were to come, the curiosity it would be possible to slake if only I could open my mind and my spirit wide enough to launch myself on the quest I knew was waiting for me.

I had heard so much by then about what the press of humanity was doing to the health of the planet. I had read widely, longed fiercely to see for myself what was happening. And why. And here, on this beastly hot night in Jordan, I was offered the chance.

And as I looked over the bank many metres below to the receding Dead Sea, my mood lifted even further. It was a hardscrabble descent of big boulders and dust, but I never doubted I could make it. I took off my fancy shoes and tucked my stockings into a pocket, hoisted my long crimson silk skirt and set off towards the water, puncturing my feet on the jagged stones as I half-climbed, half-slid, down the bank, roaring with laughter.

I could taste the sea's famed salt on my lips, smell it, feel it cake in the sweat of my skin. It was a promise. I waded in past my knees, cloaked by the night, like a guilty teenager seeking a forbidden kiss. The water was slippery with salt. Within minutes, my fingers looked like prunes and my torn feet had stopped bleeding.

Other feast-goers had made the same pilgrimage. We clung together, giddy from the heat and the sun and the salt, dancing at the water's edge, celebrating life in that ruin of a sea. Thinking back, I have come to the conclusion that it must simply be human nature to hope when none is obvious. Otherwise, why would we, some of the few people in the world to realize how terrible things were and know the urgency of fixing them, have laughed so hard as we struggled back up the bank to finish our feast? We humans are made to dream, to feel the depth of pain in the same heartbeat as we understand how to move beyond it.

I have made many journeys since the night I danced at the Dead Sea. And this book is the tale of what I found.

My story-gathering has led me all over the world. Each journey took me to a perfect example of one facet of the problem or one hint of a solution. I was near the end before I realized that I had looked for my answers on several of the world's most forgotten islands, self-contained places that have a gift for prophecy.

Some themes emerged. It is clear that humans have become as powerful a geological force as the four elements of water, air, earth and fire that the ancients believed made up the cosmos. Humans are so numerous, so ravenous, so self-centred a species that we have become the fifth element. After my travels – both physical and intellectual – I am convinced that if we do things differently we can flourish, as could other species on this planet. We're just not doing things wisely enough yet. It's understandable. We haven't needed to so urgently before. Now we must change, and I believe we can.

The other theme that became inescapable to me is that the solution to our troubles lies in acknowledging what is culturally unknowable to us at this juncture in our history as a global society. Throughout the book, I talk about this as the 'legends' of our society. By that I mean the unexamined stories that explain to us how our lives work and what our place is thought to be in the web of life. I went looking for examples of how these have changed throughout time. I found many. To me, that says that these legends are capable of changing again, with precision, with speed, with spirit. It is the human legacy to change, even as we fail to notice that we are doing so. It is our legacy – as a species and even as individuals – to keep going, even when it seems that the end has come. We may weep at the Dead Sea, but we will also learn to dance.

I decided literally overnight that I had to go to Madagascar, the impoverished island off the east coast of Africa, to write about the vanishing forests there and bear witness to predictions of a sixth extinction. I swam in the Indian Ocean at dawn there every day, went searching for lost lemurs, dodged malaria and armed troops and marvelled at how different the stars look so far below the Tropic of Capricorn.

My odyssey has taken me into deep history, to the only part of the world that can give us clues – and they are elliptical at best – about how the dinosaurs died off. I have visited the parched deserts of the Middle East, scene of the most terrible modern example of the death of an oasis. I flew to the barrens of the High Arctic, to one of the first places on the planet to have chronicled the damage done by climate change. The people who live there, the Inuvialuit of Banks Island, have learned the hard way to preserve their local resources, only to have them endangered by the activities of humans far away.

With a double-sprained ankle stuffed into my boot, I climbed a mountain of raw Precambrian shield rising out of the pristine Amazonian forests of South America, the better to marvel at one of the most extensively protected countries in the world. In Iceland, a stone's throw from the Arctic Circle, I found a people who had destroyed their country's forest cover, then found a way to unleash the power of the very workings of the inner earth in a bid to find redemption in the form of hydrogen power.

And finally, in the mythical Galapagos Islands, a part of the planet where human need does not come first and where both will and financial muscle meet to conserve nature, I explored the floor of the ocean itself and began to piece together the critical part the oceans play in this complex mosaic of life on earth.

I have, in some cases, clapped eyes on nearly every single individual in the world of some endangered species: the oryx and the onager in Jordan and the last living specimen of a whole sub-species of giant tortoise in the Galapagos, among others. At one point in the Galapagos, I saw an extraordinarily large proportion of the world's population of endangered flightless cormorants, the freakish black birds whose wings have evolved into stumps because the birds no longer need to fly.

There is something profoundly shattering about this. These are proud and marvellous creatures, exquisitely adapted. They should not be so reduced in numbers that I can spot most of them in a single afternoon.

Even worse, I came to realize that some of the most severely taxed species are also some of the most hardy. The oryx, the musk ox, the giant tortoise, the marine iguana. Each has weathered unimaginable physical deprivation over millions of years and survived. Only humans, the cleverest, the most brilliant, the most demanding of the earth's creatures, could force them to the brink.

All of it led me to Charles Darwin. Quite unexpectedly. And not to the many social adaptations of his theories of evolution and natural selection that can be made to justify any bigotry. But to the sheer biological truth of the theories that nature itself proves again and again: species are not fixed. They change to fit their surroundings; therefore, if their surroundings change, so do species.

What we are witnessing here, now, on our planet, is a profound change in our surroundings, caused by what humans are doing. So species are changing. In most cases, what this means is that they are dying out across much of their habitat because things are altering far, far faster than the leisurely pace historical evolution has prepared them for. That's how palaeontologists, trained to look at the normal pace of evolution, can identify an extinction spike today. Extinction is normal. This rate of extinction is not. It has happened just five times before in the planet's 4.6 billion years.

Nearly 150 years after Darwin first published his theories in *On the Origin of Species* in 1859, these theories are more relevant and more evident than ever. They sit at the heart of understanding the modern ecological crisis. The key is this: evolution and natural selection do not just apply to the past. That was the main hurdle for the Victorians who were faced with Darwin's theories to begin with. The challenge for modern society as well is to grasp that these theories also apply to the present and to the future.

Evolution makes up the unfailing cycle of life. Evolution, like creation, is not a one-shot deal. It does not end with us. It keeps going. The staggering rates of species threatened with extinction – 30 per cent of fishes, 25 per cent of reptiles, 24 per cent of mammals, 20 per cent of amphibians, and 12 per cent of all bird species – are the tragic evidence of that. What is the spectre of mass extinction but forced evolution? It is artificial selection of species on a wide scale, the opposite of the natural selection Darwin described. It is the planet telling us that we have gone too far.

But it's not just Darwin who has lessons for modern society. His society does too. The rooted belief of citizens of the Victorian British Empire was that the species were all formed in a single burst of creation, orchestrated by God, for the benefit of human beings. The number of species was fixed, and each was immutable. Whatever God had created in that seminal episode was still on the earth and always would be. For its part, the earth was created in 4004 BC, a figure calculated by theologians reading the Bible, not by geologists examining the earth.

So when Darwin – very reluctantly, it must be said – put forth a theory that argued that humans had just happened to evolve over time, it was profoundly disturbing, not just for the matrix of society, but also for Darwin himself. There was no dancing at the Dead Sea for him.

The rejection of his ideas rang from the highest pulpits through the halls of academe and down to the shops on the high streets. But by the time Darwin died in 1882, his theories had become the common currency of the realm. By 1900 they were contained within mainstream textbooks in Britain and North America. Within the space of a single man's life, Victorians had come to accept the theories that had originally struck at the very mythology of the human species.

More importantly, they had come to understand that their world could still go on even if Darwin were correct. They stepped back to look at the legends that informed their society, reassessed them, and fashioned them in a new way.

It's an extremely difficult task, as Robert Ornstein and Paul Ehrlich explain in their book *New World, New Mind*. As a species, *Homo sapiens* has the ability to think complexly and to discern trends but, strangely enough, the evolutionary forces that created the complex ingenuity of humans have also left us cruelly limited in our ability to understand those forces.

Ornstein and Ehrlich use a metaphor to explain our circumstances. An amoeba, placid on a microscope slide, can perceive the difference between light and dark, and can tell when it has found food, but it cannot perceive the microscope or the human observing it, much less the larger world the human operates within, or that world's past and future. We have evolved to perform in much the same way. Early examples of *Homo sapiens* survived because they were adept at perceiving danger and food in the nearby forests, not because they could tell whether the structure of the forests would be the same in a hundred years' time.

'Indeed, since little or nothing adaptive could have been done by individuals about such long-term trends, perceptions of gradual long-term change may have been actively suppressed in the course of our evolution,' Ehrlich and Ornstein point out.

And that's what is different about *Homo sapiens* today. We can change entrenched ideas, and we have. We find spaces in our legends when we need to and we are capable of using metaphor to understand change.

Ornstein and Ehrlich argue that we humans can consciously rewire our brains to think long-term. They mean that we ought to educate ourselves about the evolution-inspired tendency of our brains to operate short-term and work around it. To me, there is an even more effective alternative to doomsday. Rather than out-think our evolutionary limitations, why not invoke them?

If humans respond to caricature, use it. If humans long for a mythology to explain their society, appeal to that longing. Create a legend that humans can embrace, that they can relate to themselves, their families, their own regions and then their planet. Work with evolution, in the best Darwinian tradition.

The answer will not lie in explaining the science ever more carefully to the converted. As in Darwin's day, few people are able to judge what the Swedish scholar Alvar Ellegard called 'the possible truth or falsehood of an abstruse scientific theory'. We must bring the science to life in tales that capture the imagination, that evoke the rituals of setting impossible tasks, faltering along the way, despairing and ultimately prevailing.

Precisely four years after I celebrated in Jordan I found something that felt like proof of this theory. Drawn back to the land of Darwin's birth, I stumbled upon a remarkable cabal of people who are, in spirit and intent, dancers at the Dead Sea. They have made a new, living metaphor, an awe-inspiring garden they call Eden. Except this one is not in the fertile crescent that gave birth to the Bible's tale. Instead it lies in the Cornish landscape, cradle of the undying legend of King Arthur and of Tintagel Castle, where his conception was said to have taken place under the cover of wizardry and deceit.

Perhaps the magic still hangs in the Cornwall air. At least, so it seemed to me when I arrived, by a quirk of fate, at autumn's equinox. The Cornish lands are both blessed and cursed by geography. They are blessed because the shorelines are magnificent, the seas still generous. Plants grow here bathed in the salted breezes that are warmed by the Gulf Stream off Cornwall's coast and sheltered from the harsher winters they would have in other parts of northern Europe.

Blessed because underneath the Cornish moors lies

decomposing granite that contains something of value to modern humans: the makings of fine porcelain.

Cornwall has also been cursed. Centuries of mining has meant that the rolling hills have been horribly disfigured, the ground torn apart and the slag heaped into ghostly mounds that the locals, with their self-deprecatory pride, call the Cornish alps. The open-faced mines that yawn beside the towering spoils eventually outlive their commercial usefulness and fill with water. When the light hits them a certain way, they shimmer an otherworldly electric green.

Tim Smit, whose palindrome name sets the tone for his ability to think backwards and forwards at the same time, took one look at these barren pits and saw Eden. Since then, with the better part of £100 million of investment from both public and private sectors, he has built what amounts to the world's largest greenhouse in Bodelva, one of the abandoned pits.

He and a team of other sideways thinkers removed the water, flattened the pit's bottom, stabilized its crumbly sides with steel and grass, laid in a fertile bed of 83,000 tons of soil manufactured from recycled waste and then built some geodesic domes, called Biomes, one of which is large enough to house the Tower of London.

Then they planted. The biggest Biome contains plants that like wet heat. When you walk in, your glasses fog up. Small boys delight in stripping off their shirts and middle-aged women succumb to regret that they cannot do the same. More than 1,000 different types of tropical plants live here, from the rare to the everyday, including the endangered bottle palm from Mauritius that collects water in its trunk, the self-cloning banana plant from the tropics and the cancer-fighting periwinkle from Madagascar.

A smaller Biome houses plants from Mediterranean climates, such as olives and cork trees and all the herbs of a summer's memory. Outside are the hardier plants of Cornwall itself, including some rare and endangered heritage apple trees, and fare from every kitchen garden in England along with such tender anomalies.

Eden has been touted as one of the wonders of the world and it is, at its bare bones, one of the most successful tourist attractions in Europe, drawing about 1.3 million visitors a year, hosting music concerts and bolstering the economy of one of the poorest corners of the European Union.

As a project aimed at sheer resurrection, Eden would be cause for hope. But it is far more than that. It bears witness to the rewriting of legends. When I was there I stood just inside the entrance, watching as visitors first caught sight of the Biomes glinting blue in the regenerated pit. They registered awe, not a common emotion in this jaded world. Some visibly gave themselves over to delight, encouraged not only by the colours and smells, but by the pervading humour that makes it come alive – the ample Eve sculpted from living grasses or the steely Adam trying hopelessly to rein in his flax. It all contains a sly message: people depend on plants. And plants depend on the state of the planet. Eden is an archive of that relationship, a reminder that we have achieved everything that we have as a species largely because we were able to cultivate plants. It's a call to remember that we humans, too, have roots, and that we need them.

It doesn't end there, though. Eden has become such a success that it has spawned an international think-tank, employing people to develop ideas on everything from best rubbish practices to low-cost, energy-efficient housing. The legend of its birth and of Smit's heroic travails as Eden's midwife – the stuff of books, newspaper articles and documentaries – gives it enough muscle to induce rivals such as mining companies to sit down at the table in this revivified mine and share stories about how to mine minerals with the least social and environmental disruption possible.

As I left Cornwall, I had a revelation. When I started my journeys, I had told myself that I was looking for information, that precious modern currency. I wanted to ferret out the latest facts, sift through falsehoods, present an air-tight, neutral intellectual case.

But as I travelled back to London, and then home to Canada, I realized I had been seeking something altogether more primal: the mysterious recesses of the human mind with its ancient, elastic logic. The slippery point at which the mind begins to see things in a different way.

I was searching for meaning, for stories, for the legends that bind us together as human beings and explain us to ourselves. For something that could tell me why my species is so nonchalantly damaging the planet's life-support systems. And, therefore, how we might stop if we choose to.

To be honest, I didn't find an answer. The great joke is that there probably isn't one. But I found narrative and joy and hope. And, in the end, I think that adds up to more than the sum of all the juicy bits of knowledge I garnered.

I think this is what C. S. Lewis, author of the Narnia series, meant when he said that he was a believer not only in the efficacy of imagination and myth, but in their necessity. In his *Selected Literary Essays: Bluspels and Flalanspheres, A Semantic Nightmare*, he explained that myth is the secret ingredient in meaning. Myth carries meaning in a way that rational truth-telling cannot. 'For me,' he wrote, 'reason is the natural organ of truth; but imagination is the organ of meaning. Imagination, producing new metaphors or revivifying old, is not the cause of truth, but its condition.'

Because humans are creative. We dream. We respond to the personal. Some of us build Edens.

We have bursts of ingenuity that have taken us to the moon, to the bottom of the sea, to the magical inner workings of foetal stem cells, to the intricacies of Virgil's *Aeneid*, Shakespeare's *Macbeth* and Margaret Atwood's *Oryx and Crake*.

So if humans – you and I – choose to change how we are using the riches of the earth, the key will be persuading enough of us that we need to rewrite the legends we hold dear and then doing so by capturing imagination with narrative.

I know now that the very Darwinian endeavour of metamorphosis is possible for humans too. We are doing it – and have done it over time – with our legends, with the secret stories that stitch humanity together. Survival comes down to those stories, to the magical act of turning them imperceptibly inside out, so when scholars look back at us, 150 years on, their jaws will go slack to think that we believed we could use up the earth and live to tell the tale.

ONE

THE Devil's Chaplain

It is a bitterly cold January morning, and I am standing outside 1 Regent Street in Oxford with three months' worth of luggage, two children and no key. In fact, no prospects of a key.

Charles Darwin has lured me here, and the chance to study on a fellowship with the brilliant Oxford University ecologist Norman Myers. I have spent months negotiating to be able to come – with my employer, my husband, my programme director at Green College and innumerable over-priced housing agencies in this ancient university town. I finally found two small rooms in a student boarding house on Regent Street – at a price I can barely afford and at a crushing distance from the children's schools – and wired money sight unseen to the student I'm subletting from. That student is on holiday in France, the house's owner doesn't know I'm subletting, the other students in the house – including the one who might have my key – aren't here. And my children, who have spent all night on a plane from North America, are shivering. It's an inauspicious start.

All this to satisfy my compulsion to investigate the parallels between Darwin's Victorian age and my own. To figure out how human societies are capable of adapting the legends that help them make sense of the world. I'm convinced that my modern society is facing the same seismic challenge to legend that Darwin's society faced in 1859 when he published *On the Origin of Species*, that, in some ways, the modern debate about ecological crisis still revolves around the place of humanity in creation. Perhaps it is even still about evolution. And I think that if I come to understand what the complex debate in Victorian society was really about, I might be able to gain insight into what is really happening in my own society.

My constant companion for months now has been Darwin's diary of the five-year journey he took in the 1830s as the amateur naturalist aboard the HMS *Beagle*. *On the Origin of Species*, which I have also been immersed in, is so carefully written as to be ponderous. But the diary is a personal, often uncomfortable, sometimes beautifully written chronicle about lands unimagined, animals unseen until then. I read it over and over, attaching neon orange notes to the pages I can't bear to leave behind.

My fascination comes from travelling around the globe with Darwin, but also from soaking up the clues he left about the social values and practices of his time and tracking how these have changed. For example, his odyssey started with the flogging of several of the *Beagle*'s crew the day after the ship set sail because the men had become drunk and 'insolent' three days earlier on Christmas Day. Not quite the labour practice of today.

In Brazil, Darwin could not hold back his disdain for the locals: 'as contemptible in their minds as their persons are miserable. . . . But the prospect of wild forests tenanted by beautiful birds, Monkeys & Sloths, & Lakes by Cavies & Alligators, will make any naturalist lick the dust even from the foot of a Brazilian.'

His horror of the slave trade that flourished there in 1832 was profound, a contrast to the attitudes of most of his companions, who saw slavery as an intelligent economic system. Darwin told of riding past a steep hill of granite near Rio de Janeiro where a group of escaped slaves lived until they were captured by soldiers. One old woman 'dashed herself to pieces from the very summit' rather than be enslaved again. 'I suppose in a Roman matron this would be called noble patriotism, in a negress it is called brutal obstinacy!' Darwin wrote, adding later: '. . . how weak are the arguments of those who maintain that slavery is a tolerable evil!'

And Darwin was convinced that Europeans – the British in particular – were more advanced than humans in other parts of the world. 'The Tropics appear the natural birthplace of the human race; but the mind, like many of its fruits, seems in a foreign clime to reach its greatest perfection,' he wrote from Rio de Janeiro in May 1832.

I'm jolted back to the England of today by the cold and by the dilemma of the key. We stop everyone who passes by, asking for help. Eventually a woman who lives further down Regent Street lets us drag our luggage into her spare room, an act of faith on both our parts. We warm up with hot chocolate and greasy sandwiches at a café, and by the time we come back a few hours later, one of our new roommates is there with a key.

I BEGIN TO FEEL as though I live at the Radcliffe Science Library. Like the magnificent Bodleian, where I also spend time, the Radcliffe has an ancient system. You request books – the concession to modernity is that you can sometimes do this by computer – and someone unseen scrambles to hidden chambers and retrieves them for you in a day or two. Taking them out of the building is forbidden, so I spend hour upon glorious hour reading them at wide, thick wooden tables next to leaded glass windows, taking notes, learning the epic tales of Darwin.

Charles Darwin was an unlikely adventurer. He ended up on board the *Beagle* only after a timid test of wills with his father, Robert, a well-to-do physician who was convinced Charles would amount to nothing and told him so. Charles had drifted through classes at the University of Edinburgh and then Cambridge University, interested only in collecting beetles and running up debts that his exasperated father was loath to pay. Finally, Robert Darwin ruled that Charles was to be a clergyman. It was a respectable calling for a gentleman, and Charles agreed, reckoning that he could write sermons and still collect all the beetles he wanted in his spare time.

When Charles got invited to go on the *Beagle*, his father told him it would be a 'useless undertaking' and said he couldn't go. Charles enlisted the help of his uncle, Josiah Wedgwood, of Wedgwood china, and together they took on the elder Darwin's objections. The turning point came when Charles reminded his father meekly that he would have no occasion to spend money while travelling the wilds of the world. His father finally consented, and Charles went to London to prepare for the journey. Two months later, after a farewell luncheon of mutton chops and champagne, Darwin set off on the *Beagle* with a fair east wind to be a companion to its captain and take a few notes on the geology and nature of the lands he hoped to see.

*

DARWIN'S RELUCTANCE to challenge his father over the journey on the *Beagle* was a sign of the heroic struggles he would face later as he developed his controversial theories. In the best poetic tradition, Darwin's struggle was not just with his society, but, more painfully, with himself.

He was not looking for a new way to explain the great mystery of how life came to be. He left England convinced that he knew that already. Like almost everyone else in Western society, belief in a Christian God was the matrix of his life. A central part of the belief was that God had fashioned the world and all forms of life in a single, brilliant burst of creation a mere six thousand years earlier. Each species was unchangeable and everlasting. Extinction of a species was equally as impossible as its adaptation: God had made exactly the plants and animals that were needed and would always be needed. This idea was called the 'fixity' of species by the Victorians. In fact, when Victorians looked around and saw the marvellous sights, tastes, textures and smells of an abundant nature, they saw all this as proof of the munificence of God.

This view of life's origins was inextricably woven into several of the central tenets of Christian theology as they were then understood. The biblical story of the creation of Adam and Eve and the Garden of Eden holds within it the ideas of original sin, the fall, atonement and redemption. And each of these was infused with the understanding that human beings were God's highest and best creation, made in His image, assured of a final reward. Other creatures were there to help make life pleasant for men and women. To Victorians and many generations of Christians before them, the idea that God's guiding mind was behind the workings of the world gave intrinsic meaning and purpose to life.

Darwin's task on the *Beagle* was to bolster the fixity theory. Instead, the trip fatally undermined it, proposing that species changed to adapt to their surroundings and that all living things had evolved slowly over time from a few simple organisms.

That meant that random factors, such as the freak extinction billions of years ago of a dozen-celled marine creature, or aimless changes in the environment a billion years after that, had led to the assembly of creatures on earth today. Including humans. And that torpedoed the idea that humans are on earth because we have a divine purpose or a special destiny. It was unbearable, and not only to Darwin's society, but to Darwin himself. What if humans were just another transient life form? What if humanity, too, would eventually be forced into extinction by a capricious nature?

The glimmerings of this idea came to Darwin while he was visiting the Galapagos archipelago near the equator in the Pacific Ocean. These are islands born of volcanic fire deep within the mantle of the earth. They first emerged from the sea as molten, naked land masses devoid of life. A few plants and animals gradually colonized these volcanic tips, coming mainly from the South American continent about 1,000 kilometres away. Most species that attempted the journey died on the way. That's why the islands have no natural amphibians, which are relatively fragile, and just four native land mammals. Instead, the archipelago is the playground of reptiles, the hardiest of animals. Those that reached the bare lava islands had a perilous time. Just a handful of species took root – cacti, iguanas, tortoises, sea birds and finches among them – in one of the most pared-down ecosystems in the world.

Darwin was resolutely unimpressed when he got there in 1835. Naked black lava soaked up the sun's unbearable heat like a sponge. Water was scarce. The stunted plants stank. It was, Darwin wrote, like being in the bowels of hell.

He spent just over a month on these strange scorched islands on the equator, choking back cinders and dust.

'I should think it would be difficult to find in the intertropical latitudes a piece of land 75 miles long, so entirely useless to man or the larger animals,' he wrote of Albemarle, the largest of the islands, now known as Isabela.

It was the giant land tortoises of Galapagos that first twigged Darwin's attention. Like other ocean voyagers of the day, he saw the giants mostly as a source of meat, and the islands as a handy fast-food market in the middle of the ocean. In his journal he remarks that the babies are tastier than the adults.

'Young tortoises make capital soup – otherwise the meat is but, to my taste, indifferent food,' he wrote.

Yet he also noted in his ornithological writings that the inhabitants of the Galapagos could tell on sight which island a tortoise had lived on by the markings and shape of its shell and body. He couldn't understand it. Why, if the islands were all roughly the same and had roughly the same parts or niches of nature to fill, would the tortoises be different? What was the sense in that? Could they simply be varieties of the same species? It was a heretical thought. That would imply, he wrote, that species could alter to fit their surroundings, and did. It might even imply that creation did not begin and end with God but was a continuous force of life.

'If there is the slightest foundation for these remarks, the Zoology of Archipelagoes will be well worth examining; for such fact would undermine the stability of species,' he wrote.

Inextricably bound with this idea, though, was the anguishing implication that God's plan – if there were one – was not as advertised at the pulpits of the day. Life could not have been created in that orderly godly burst that had formed the Garden of Eden. Instead, the process of creation was chaotic, constant and subject to chance. And in that case, how were Victorians to understand the concepts of original sin, atonement and redemption? What was the purpose of life?

The proof that Darwin needed for his scandalous theory of natural selection came from a group of finches he collected on the Galapagos Islands that would eventually carry his name. While he was on the islands, though, these birds barely penetrated his consciousness. He collected them haphazardly, not routinely marking which island in the archipelago each was from, because he thought it couldn't matter. He didn't even realize they were all finches. But when he got back to England, he gave the Galapagos birds to the eminent ornithologist and taxidermist John Gould, who realized within six days of receiving them that each was an unusual sort of ground-living finch that looked different from the others.

Within three years of returning from his journey with tales of the unimagined, Darwin had become convinced that species were not fixed. But he was not prepared to play the subversive. It was quite the opposite. He told one of his friends that it was 'like confessing a murder' even to piece his theory together. To another, he said the theory opened him to 'reproach' for being a 'complete fool'. He called himself the 'Devil's chaplain', a rebuke to the vocation as a clergyman his father had wanted for him.

Darwin refused to publish his theory for twenty-four

years after he left the Galapagos Islands, spending his time instead wrestling with his guilt, amassing more and more evidence for his ungodly theories and trying to make his own reputation bulletproof.

WHEN I COME TO KNOW Norman Myers, who is an internationally celebrated environmental theorist, I recognize some Darwinist traits in him. There is the wanderlust, the mental muscle, the intellectual imperative to search for answers, even when it's not convenient, to questions others have not asked. Myers, though, is not burdened by the crippling fear that dogged Darwin's life. He has been an environmental scholar and activist for nearly half a century, ever since the giraffes and zebras of the Kenyan savannahs first fired his imagination. Myers was the first to understand that the sixth mass extinction in the history of the planet is under way and that humans are its agent. He was the first to add up the billions of dollars a year that governments are giving to subsidize industries so that they can destroy the life-support systems of the planet. He is the one who figured out that as the warm, wet parts of earth around the equator are being ruined so are the powerhouses of future evolution. And he has helped form a conservation strategy to stave off extinctions based on the medical model of triage: save the most important first. That involves identifying the 'hotspots', places like Madagascar, where many unique species are poorly protected.

Myers has paid the price by being considered a maverick, even among biologists. He has been vilified with regularity over the decades by colleagues and foes. And he has won international acclaim for his work, including the prestigious Blue Planet Prize in 2001 and an appointment as companion to the Order of St Michael and St George by Queen Elizabeth II in 1997 for 'services to the global environment'.

But he's wary. He rarely takes students. He is closing in on seventy and is not at the stage of life to suffer fools gladly, if he ever was. By some miracle, though, he has agreed to hear me out, and today I have an appointment to meet him at his home office up an obscure lane in Headington, a twenty-minute taxi ride from my college.

He is welcoming but frail, just recovering from cancer and woefully busy. My allotted sixty-minute appointment with him must do double duty with his lunch. He potters gingerly around the kitchen among containers of wan amaryllis plants stretching for the bleak January sun. He heats himself up a bowl of soup and starts timing the meeting the second we sit down. I talk and talk, telling him about my fascination with Darwin, my conviction that there may be parallels between the Victorian age and our own, my need to understand why our society is in such denial about the consequences of ecological disaster.

By the time I'm finished, he's smiling and his eyes

are on fire. He raps off a list of books and articles I ought to read – most of which he offers to lend me – and questions I might think of asking. And by the time our sixty minutes are up, I am lugging a backpack full of his books to my tiny room in the boarding house on Regent Street before dashing to pick up my children from their English schools.

DARWIN WAS FINALLY prodded to publish his controversial theories of evolution and natural selection. Another travelling naturalist, Alfred Russel Wallace, had developed similar ideas from his work in the Malay Archipelago and sent a paper on them to an astonished Darwin for critique. Eventually, Darwin and Wallace published their separate papers at the same time in 1858 and Darwin raced to push out *On the Origin of Species* the following year. The volume caused a furore, even though the theory of evolution, or adaptation of species over time, had been brewing for decades. The extra piece Darwin added – the theory of natural selection – was even more controversial. He proposed that all living things are involved in a struggle for survival. Some do better in that struggle than others and so pass down the winning characteristics. Over time, these characteristics become more common.

On the Origin of Species was an instant best-seller. The 1,250 volumes 'bound in royal green cloth with fine gilt on the spine' sold out the day they were published, 24 November 1859.

A second edition of 3,000 was rushed into print for the following January. The sales baffled Darwin's publisher, who had adjured him simply to write about domesticated pigeons and leave controversy alone. Pigeons, the publisher assured Darwin, were a safe sale.

Thomas Henry Huxley, the brilliant scientist who became Darwin's most vociferous defender, jumped at the chance to write an early and influential review. It was resoundingly positive. Darwin himself, pleading the never diagnosed ill health that would plague him for the rest of his life, refused to leave his home in the countryside of Kent to defend his ideas.

He would write. He would – and did – revise his writings extensively. But he would not debate in public. That he left to the vigorous Huxley, who excelled at repartee and had a talent for attracting attention.

By hiding from sight after his heroic world travels, Darwin became all the more intriguing to the romantic Victorians, who saw him as a virtuous recluse with a mysterious past that had been full of danger. As well, both Darwin and Huxley, the two public faces of the theory, were eminently respectable gentlemen, educated, the right class and scandal-free. That made it harder to dismiss the theories they put forward.

Sir Julian Huxley, the grandson of Thomas Henry, wrote: '. . . Darwin was perhaps the most respectable revolutionary the world has known.'

Still, the debate was ferocious. But this was before the discovery of genes and genetic inheritance. Before scientists discovered that genes mutate and that evolution takes place in the very molecules of living beings. It was even before the field of palaeontology was well established. The eminent scientists of Darwin's day were perturbed by the fossils of large and clearly very ancient creatures that were being discovered as close to home as Oxford and throughout Europe. In 1842 Richard Owens, a professor at the Royal College of Surgeons in London – and later a fierce anti-Darwinist – coined the term 'dinosaur' as he struggled to describe a megalosaurus bone found just north of Oxford.

Now we know that that very fossil is 168 million years old and from the Jurassic period. And that the megalosaurus is extinct, along with roughly 99 per cent of all the species that have ever existed.

But Darwin didn't understand any of that. Nor did other scientists or the public at large. So Darwin's theory was not being challenged on scientific grounds. This was not, in fact, a scientific debate. What was it really about, then? It is only when I begin to mine the Radcliffe's wealth of analysis that took place to mark the 100th anniversary of the publication of *On the Origin of Species* that I finally begin to understand what was happening in Darwin's day. One hundred years on, the things that had been culturally unknowable no longer were. Scholars were able to identify the legends that lay underneath belief.

The most persuasive piece is a painstaking analysis by the Swedish scholar Alvar Ellegard, who looked at the reaction to Darwin's theories in the British periodical press shortly after *On the Origin of Species* appeared. This includes the magazines and newsletters that the general public would have read. Ellegard concluded that the debate was about the Church's need to hold on to its authority, dressed in the clothing of science. The fight took place on the level of metaphor. It was not about true or false, or about good science or bad. This was a debate about the fundamentals of good versus evil.

It was this unsophisticated debate that eventually filtered from the Anglican and academic establishments through to the middle class and finally to the uneducated public. And it was this debate that threatened to suppress Darwin's theories.

Ellegard quoted the Roman Catholic *Dublin Review*, published around the time *On the Origin of Species* came out, which summed up the establishment view: 'The salvation of man is a far higher object than the progress of science: and we have no hesitation in maintaining that if in the judgment of the Church the promulgation of any scientific truth was more likely to hinder man's salvation than to promote it, she would not only be justified in her efforts to suppress it, but it would be her bounden duty to do her utmost to suppress it.'

The bid to suppress was all the fiercer because the Church had been forced to give way on earlier scientific questions. The Church had savagely fought the findings of Copernicus and Galileo that the earth revolved around the sun.

When Galileo published a book in 1632 asserting that the earth was not fixed or at the centre of the universe, he was arrested and tried by the Inquisition. Found guilty and sentenced to life imprisonment, Galileo recanted. The Church banned his book 'as an example for others to abstain from delinquencies of this sort'.

But when Galileo's scientific findings were broadcast anyway, they began to raise widespread doubts about the teachings of the Church. As they took hold, the Church fathers ended up being forced to comfort the public by asserting that the important lesson in the Book of Genesis was not so much the account of the earth's creation, but man's.

Then came Darwin. Take away the biblical story of the creation of Adam and Eve, and it was feared that the centre of Christianity could not hold. As Ellegard wrote: '. . . the spiritual and moral welfare of the community was a more important consideration than the freedom of scientists to divulge their theoretical conclusions'.

It was only on 23 October 1996, almost exactly 137 years after the publication of *On the Origin of Species*, that Pope John Paul II finally embraced evolution. Darwin's theories, given life on the hellish islands of the Galapagos, just make sense, said the Pope, although, he added, evolution has nothing to do with the human spirit.

'It is indeed remarkable that this theory has been progressively accepted by researchers, following a series of discoveries in various fields of knowledge,' the Pope said in his Message on Evolution to the Pontifical Academy of Sciences. 'The convergence, neither sought nor fabricated, of the results of work that was conducted independently is in itself a significant argument in favour of this theory.'

Today, because the mainstream understanding of Christianity is quite different from that in Victorian times, it's difficult to grasp just how agonizing it was to the public to contemplate a world not organized by God and set up for the benefit of humans. Darwin's theories pricked human pride because they posited that we are a come-by-chance species rather than the final and deliberate act of an all-powerful God. Embracing Darwin's theories just because they made sense was not nearly a strong enough reason to abandon a system of faith that explained so satisfactorily how important humanity was. So resistance to Darwinism ran wide and deep, from the highest pulpits right down to the High Street. This was an attack on the very mythology of how humans had come to be. The low-brow mass-market periodical *Family Herald* put it this way in its issue of 20 May 1871: 'Society must fall to pieces if Darwinism be true.'

Darwin himself was not prosecuted for publishing his theories, although his chronic mystery illness meant that he was, in a sense, incarcerated in his home. But others were taken to task for supporting him. The Oxford geometry professor Revd Baden Powell and several other authors, writing in a church manifesto *Essays and Reviews*, supported Darwin, saying that nature evolved on its own. They were accused of heresy, and two were prosecuted.

Nevertheless, by the 1870s the debate over evolution was largely settled among the educated men and women of the Western world, although a movement to discredit the theory in favour of the idea that God created all creatures still thrives in many parts of the United States. By 1871, Darwin finally had the courage to publish *The Descent of Man*, which laid out his scandalous theory that humans and apes have a common ancestor. By the time he died in 1882, Darwin was hailed as a national treasure, buried in Westminster Abbey next to Britain's monarchs and mourned as a genius. By the close of the century, high-school textbooks in zoology, botany and geology in both Great Britain and the United States carried sections on evolution.

Despite the skirmishes mounted by creationists – including the Scopes trial in Dayton, Tennessee, in July 1925, which was covered by more than two thousand newspapers and radio stations, and the sporadic attempts even today in Britain and the United States to ban the teaching of evolution – Darwin's once-controversial biological theory has become so mainstream as to be unremarkable.

As we approach 150 years since the publication of *On the Origin of Species*, Darwin's ideas are so much part of the fabric of thought on how life came to be that they have faded into the background, seemingly no longer in need of defence.

And what of the Victorian belief system that Darwin's twin theories unseated? The vast majority of Christians have, like the Pope, found a way to swallow their religious faith and evolution easily in the same draught. Despite the naysayers of the late 1800s, society is no more fractured today than it was before Darwin published his theories. Over time, the legends that explain the human place in nature have adapted. Even evolved.

Yet when I look around me at the abundance of species threatened with extinction, at the receding forests, the damaged climate, the fouled waters, I wonder if humans really have come to understand and accept our place in nature. Yes, we understand that we evolved over billions of years, along with other species.

We understand that most of the species that ever lived have become extinct. Like the dinosaurs, their shelf life has expired. But do we understand that species continue to evolve? That evolution does not just apply to the past, but must also apply to the future? That *Homo sapiens* is not necessarily the highest and best evolution can do?

This is Darwin Round Two, 150 years on. Our modern society is now being called upon – by the imminent threat to the planet and to our very existence as a species – to take the next step in embracing Darwin's revolutionary theories. As before, this will be an affront to human pride. In the end, we will come to understand that all creation is not here for us. We are but one species – although an extraordinarily powerful and intelligent one – dependent on other species and planetary life-support systems for our own existence. *Homo sapiens*, too, has a shelf life. And rather than trying to prolong it, our actions as a species are aimed at shortening it. Not the immediate shelf life of individuals, your neighbours and mine, but of the species as a whole.

OXFORD IS A FINE ARCHIVE of other brave battles over human values that have been fought, won and lost over time. My children and I have scrounged rain gear from Oxfam charity shops and bought bikes to get around. Every school day we form a convoy – I am like a mother duck trailing her two sodden young behind – to ride and then walk through the winter rain for about two hours each way to get all of us to school. It is a glorious, contemplative time. Our route takes us across ancient cobblestones, down a lane haunted by soldiers from the English civil war and around the neo-Gothic monument to the three Protestant Oxford martyrs, burned at the stake in 1555 – not so very long ago, after all – for holding their own against 'the errors of the Church of Rome'.

We ride past the colleges that have moulded many of the world's eminent thinkers, politicians and religious leaders, some established as early as the thirteenth century. Past the early seventeenth-century Bodleian Library, founded by Thomas Bodley, an Englishman who grew up in Genoa because his Protestant family was forced to flee England and religious persecution under the reign of Mary I. But who had the vision to make sure when he returned that students could open their minds with fine buildings and books. Now Catholics and Protestants, not to mention atheists, Jews, Muslims and those of many other systems of belief, dream in these halls.

SO MANY WAYS of thinking can change, and do. So many of the unexamined stories that tell humans how their world ought to work gradually come to be re-written. These stories are by nature shifty, inchoate, plastic, it seems to me, and therefore subject to reform. Jesus College, which we pass by on our route, always gives me a chuckle. It was founded in 1571 by Elizabeth I, the child repudiated by her father because she was not a boy, and who went on to invent herself as the powerful virgin queen. Just over four hundred years later, in 1974 – not so very long ago either – Jesus caused a sensation in Oxford by becoming the first all-male college to let women in. The others have reluctantly followed suit. I grimace when I think of my own programme, which is welcoming to women – as long as they are not mothers. I discovered after I got here that the programme has lodgings to rent to its students, including rooms in the mansion on Norham Gardens that houses the programme office. The mansion is just a few minutes' ride from my children's schools, which we cannot fail to notice each day as we do our marathon commute, but the rooms are by fiat not rented to scholars with children. This is not a hidden intolerance. At one point my programme director, a wonderful documentary-maker with a notably high social conscience, takes me aside. 'Pity you had to bring the children,' he says. Late one night, I let my son run his shiny red motorized race car through the halls of the splendid mansion while I work.

As I read and think and let my mind loose at the grand wooden tables in the Radcliffe Library, I realize that it's not so much that *Homo sapiens* is doing anything new or different. It's that there are so many of us now – 6 billion and rising by about 72 million a year. Practices that have been commonplace for millennia, such as cutting down trees and hunting animals for food and fishing the seas, have become unsustainable because so much of it is happening at once. There are no new bits of the planet to take over. Even the seas are being colonized. And despite the mounting scientific assessment that these practices are dangerous, little is being done to stop them. In fact, the pressures to keep going further down the same road are mounting.

Darwin would tell us it can't work for long. Change the living conditions greatly and life will change, probably in unexpected ways. It's as if the evolutionary forces of the planet, sedate and gracious in the main, are winding up, poised to spring into merry chaos once the fiddling goes too far.

Nature doesn't take prisoners under such circumstances. The fossil record shows us how this has played out during the planet's five earlier mass extinctions. During the most extensive, the Permian extinction 250 million years ago, 54 per cent of families – branches in the tree of life – were wiped out.

Unlike in Darwin's day, the science today is not just theoretical. It is tangible, visible, quantifiable. We can count the number of species on the brink of extinction, witness whole ecosystems crash, see forests slashed and burned. But an especially clear example is climate change.

Since about 1990, the scientific investigation on climate change has been strong enough to show that there's something to worry about. This means that global climate systems at some point in the future will undergo massive change as a result of the high concentrations of carbon dioxide and other greenhouse gases that humans are pumping into the atmosphere. And there is strong evidence that this massive climate change is already happening.

Climate is an exquisitely finicky thing to predict. Even climatologists will tell you that. It entails using supercomputers and making a lot of assumptions. There are all sorts of natural variations in the weather. And the climate itself – climate is what you expect and weather is what you get – goes through large-scale changes of its own accord over millennia without the intervention of humans.

Even when all this finickiness is taken into account, an overwhelming majority of the planet's most distinguished scientific experts on climate hold to the theory that big changes in the climate are definitely happening. Perhaps catastrophic ones. The consensus is so large now that it is as near unanimous as scientists get.

By February 1992 the Royal Society and the National Academy of Sciences, two of the largest scientific organizations in the world, had put out a sobering message, an unprecedented consensus of the two world science powers. It is a warning about shelf life. Our own and that of other species.

'Unrestrained resource consumption for energy production and other uses, especially if the developing world strives to achieve living standards based on the same levels of consumption as the developed world, could lead to catastrophic outcomes for the global environment. Some of the environmental changes may produce irreversible damage to the earth's capacity to sustain life. Many species have already disappeared, and many more are destined to do so. Man's own prospects for achieving satisfactory living standards are threatened by environmental deterioration, especially in the poorest countries where economic activities are most heavily dependent upon the quality of natural resources.'

Just months later, on 18 November 1992, an impressive 1,500 of the world's top scientists from sixty-nine countries – including a majority of the living Nobel laureates in the sciences – issued a warning to humanity on the same topic. These representatives of the Union of Concerned Scientists implored humans to act in their own self-interest.

'Our massive tampering with the world's interdependent web of life – coupled with the environmental damage inflicted by deforestation, species loss and climate change – could trigger widespread adverse effects, including unpredictable collapses of critical biological systems whose interactions and dynamics we only imperfectly understand. Uncertainty over the extent of these effects cannot excuse complacency or delay in facing the threat.'

This is Darwin distilled and applied to the future. The same scientists, a little further in their statement, went on to say this: 'We must recognize the earth's limited capacity to care for us.'

It might be Darwin trying to explain to his Victorian audience that the world was not created for the benefit of humans and the world's life forms are not fixed. Now, modern scientists are discovering that ecosystems and global life-support systems are not permanent. There is no fixity of the ecosystem, to put it in Victorian terms. The earth is not inexhaustible. Life forms will adapt to change, and not necessarily in ways we want.

More recently, in January 2001, the 2,500 scientists and other specialists hand-picked by world governments to sit on the Intergovernmental Panel on Climate Change reported from China that the change in the average global temperature could rise by as much as 5.8°C over the next century. That's more than the rise in temperature that banished the Ice Age ten thousand years ago. World politicians have responded to the threat by signing the Kyoto Protocol, which aims to reduce greenhouse gas emissions in some industrialized countries to below 1990 levels. That's a tiny fraction of the reductions that would be needed to stabilize the climate.

Yet on the surface, there appears to still be debate over the science of climate change. US president George W. Bush, whose country is the biggest emitter of greenhouse gases in the world, continues to question the science's validity. One of his first acts after gaining office was to pull out of the Kyoto Protocol, saying the science is unproven and the Kyoto targets for cutting down carbon dioxide pollution may not even work. The world scientific community took him to task for this, saying that his strategy looked suspiciously like foot-dragging. The science, they said, is already clear, if inconvenient.

Shortly after, the oil and gas producer Exxon Mobil, the most profitable corporation in history and a big donor to Bush's election campaign, began calling for better science too. Until the science gets 'crisp', a company official said, there's no sense making policy. 'Science is a process of enquiry,' explained Frank Sprow, Exxon's head of safety and environmental health. 'I'd like the answer tomorrow afternoon, but it may be a decade before the science really gets crisp, because there's so much fundamental information that has to be worked on.'

You would be forgiven for believing that the climate-change debate is about science. But it isn't. That is just an alibi. The climate-change debate is really about symbols, just as the debate over evolution was in Darwin's day. The climate-change debate pits good against evil, stacking an old, familiar economic orthodoxy that is manifestly toxic to the earth against the idea of a new system that could be environmentally benign.

In fact, replace some of the theological ideas in the Victorian quotes with modern economic concepts, and you come up with exactly what the modern defenders of unsustainable development are saying.

Here is Ari Fleischer, then press secretary to the president of the United States, the richest country in the world, on 28 March 2001: 'The president has been unequivocal. He does not support the Kyoto treaty. It . . . is not in the United States' economic best interest.'

And President George W. Bush himself the following day in Washington: 'I will explain as clearly as I can, today and every other chance I get, that we will not do anything that harms our economy. Because first things first are the people who live in America. That's my priority. I'm worried about the economy. . . . And the idea of placing caps on CO2 [carbon dioxide, a greenhouse gas] does not make economic sense for America.'

It's an eerie echo of the Roman Catholic *Dublin Review*'s fierce rejection of Darwinism in the 1850s, which said, roughly speaking, that the highest purpose of man is the salvation of man. If that means suppressing science, we must suppress science. President Bush is saying that the highest purpose of man is to maintain and increase the standard of living for the planet's wealthiest societies in the short term. If that means damaging the planet in the long term, and maybe even harming humanity's long-term standard of living, then we must damage the planet.

There's an irony in all this that would be pleasing if it weren't so tragic. Some of the most eminent economists in the world who are affiliated with the Union of Concerned Scientists, including two Nobel laureates, have concluded that cutting down on greenhouse gas emissions would actually save money and preserve living standards.

The corollary is that not cutting down on emissions is bound to harm world economies very seriously over time. Logically, then, if the goal is to preserve an idealized standard of living, the route is to go all out cutting emissions, at least according to one eminent school of thought.

A paragraph I stumbled on in the Radcliffe keeps playing in my head as I try to sort this out: 'History to be true must condescend to speak the language of legend; the belief of the times is part of the record of the times; and though there may occur what may baffle its more calm and searching philosophy, it must not disdain that which was the primal, almost universal motive of human life.' It's from H. H. Milman in his 1855 treatise on the *History of Latin Christianity*. It strikes me that this is the key to understanding why we're not doing more to help fix the ecological crises of the planet when we could.

It's about the legends of a society. Some are very local, like the ones that told the men's colleges of Oxford to keep women out and confine them to their own, less lustrous Oxonian colleges, even as other universities all over the world were going coed. Other legends are universal, like the belief that the earth will keep providing no matter how we stretch its means. The trick is to figure out what these legends are, so we can figure out what the underlying, unstated value system is. This is the seminal thread that links humans together, and it is far more powerful than mere information.

And it's the ideas that threaten legends that are the hardest to explain in terms of information. Look at other seismic shifts in thought. Galileo and his theory that the earth revolved around the sun. He was excommunicated for it because it kicked up against the biblical teachings of the day that said God had formed the earth and set it at the centre of the universe. Darwin, who believed that God had not assembled all life's plants and animals in a single burst of creation but that species evolve over time as they adapt to new surroundings. This too was heresy. Worse, it was abstract heresy. There was, then, no proof, not even the understanding of the mechanism of genes or heredity. Einstein's theory of relativity was not nearly so heretical, nor is the modern research on the human genome. They are powerful ideas, but they do not put at risk the legend of how humans fit into the world.

I MAKE THE TREK back to Myers through a lane now filled with spring flowers. He has been in frequent, brief contact, sending me newspaper clippings by letter and book references by email, and phoning me with inspiration. He has colour in his face today, bounce in his step. His amaryllises are tall and strong in their pots. The cancer is gone. We pass through the kitchen and into his sitting room, with its fat sofas and textiles from lands far away. I have been tortured with a question I want to ask him. I am afraid of his answer, but plunge in anyway. Are humans a suicidal species?

No, he replies, to my amazement. Humans are capable of planning, of changing behaviour at the flick of a switch if they are persuaded they need to. They are not by nature mean or even self-centred. Even governments have invested in long-term social goods like health and education, despite the fact that such spending doesn't give payoffs for a decade or two. Political leaders have grappled with the vanishing ozone layer and put that part of the atmosphere on the path to recovery. Citizens have forced apartheid to fall in South Africa and institutionalized bigotry to recede in the United States. In Thailand, families had an average of seven children each just twenty years ago. Today, that average is two.

The problem is inertia, not inability, he says. People are doing things the way they always have, content to believe that things won't change. They don't know that they could have a much higher standard of living for longer if they used the planet's resources sustainably. They don't understand that the ability of nature to keep providing will falter if they don't.

As we part, he gives me another wicked smile. Keep going, he tells me.

I GO BACK to Darwin's journal. Near the beginning of his long voyage, he set foot on the very southern tip of South America at Tierra del Fuego. Part of the purpose of the *Beagle*'s voyage was to return several local Fuegians that the captain, Robert FitzRoy, had captured as hostages in 1830 as he tried to map that part of the world. FitzRoy ended up taking them back to England to teach them good English values. His goal was to return them to Tierra del Fuego so they could pass on the benefits of modern civilization, including a European garden and some proper houses, to their fellow Fuegians.

Darwin and the others mistakenly called the local indigenous peoples the Tekenika. In fact, though, they are the Yamana. The misunderstanding arose, scholars have since discovered, from the Yamana phrase for 'I do not understand you', which was 'teke uneka'.

Darwin described the scene when one of the transplanted Yamana, Jemmy Button (named that way because FitzRoy had purchased him in exchange for a mother-of-pearl button), arrived back home to his family of naked, paint-streaked warriors. Jemmy had been away three years and was by then used to a waistcoat and cravat and the rituals of an English school.

'We were sorry to find that Jemmy had quite forgotten his language, that is as far as talking, he could however understand a little of what was said. It was pitiable, but laughable, to hear him talk to his brother in English & ask him in Spanish whether he understood it,' Darwin wrote, a spooky historical note on the ephemeral nature of one's mother tongue, or perhaps a misunderstanding of Jemmy's reluctance to speak his native language in front of his English masters.

As the *Beagle* departed, Darwin wrote: 'I am afraid whatever other ends their excursion to England produced, it will not be conducive to their happiness. They have far too much sense not to see the vast superiority of civilized over uncivilized habits; & yet I am afraid to the latter they must return.'

Still, Darwin and FitzRoy harboured hopes that the civilization of the Fuegians would take hold and alter the habits of the 'truly savage inhabitants'. In the end, it didn't work out that way. Partly as a result of being exposed to the diseases of the Europeans, the Yamana people have since died out, leaving little trace beyond Darwin's observations.

EVERY MORNING this winter, as I have been following Darwin through his journals across the sooty lava of the mythological Galapagos Islands and through the drenched rainforests of Brazil, I have also been following my young son to his school through an English lane.

My son and I have walked and watched as the weather has warmed and the shrubs and hedges have come back to life. We have cursed and marvelled at the ceaseless sheets of rain that turn everything glistening green. Now, spring has come and the scent of blossoms fills the air in the lane.

All term, while I have been puzzling over Darwin, my son has been learning modern English values about the people of South America. He can talk knowledgeably now about the indigenous Yanomamo people and the complex Amazonian rainforest they live in. The Yanomamo, a South American tribe not so far different from Darwin's Yamana, are one of the least modern people in the world, first discovered living naked in the Amazonian jungle by outsiders in the 1960s. Since then, these once obscure 23,000 people have become international symbols of unspoiled nature and of the endangered rainforest. Their story has spawned a best-selling anthropology text, films, news stories, magazine articles and a traveling eco-opera.

The aim of all this international attention is both to preserve the forest where the Yanomamo live and to prevent their unique culture from being tainted by the outside world. It's a strikingly different aim from that of Darwin and the *Beagle*'s crew just over a century earlier.

And it's not completely successful. The Yanomamo are already being affected by diseases from far away; some anthropologists fear that the tribe will eventually disappear despite the efforts to save it.

As they have learned about the Yanomamo, my son and his classmates have created a rainforest in their schoolroom, complete with an understorey, liana vines, a canopy and lots of jungle wildlife. The harpy eagle, the jaguar and the strangling fig are real to them. The Amazon River has a powerful hold on their imaginations.

It's an object lesson in how legends evolve. The legend that drove Darwin and FitzRoy to hope for the 'civilization' of the Yamana has vanished from the mainstream. My son and his English schoolmates have absorbed instead a legend that tells them to cherish indigenous peoples and leave them alone.

Not to mention the ecosystem that keeps them alive. This week, on one of our final walks, my son did a reprise for me of one of his favourite songs about the Yanomamo.

It was about the tree of life. And as he hopped through the puddles, his young, thin voice rang out over the fields, telling anyone who would listen that life is all connected, beginning and end.

It seemed to me just then that I could draw a straight line from Darwin, scrambling over the blazing lava of the Galapagos, a desperately reluctant revolutionary, to my damp son, singing blithely about saving tropical ecosystems in an English lane.

TWO

IN Search of Lemurs

Madagascar feels most ancient at night, after the urgency of the daytime heat has lifted. You sense that you are in a place where time has moved more slowly than in the rest of the world. You are conscious of being on a fiercely isolated island where evolution has taken an unearthly course. The evidence is all around, even in the dark, in unfamiliar sounds and smells and darts of light. The plants, the birds, the bugs, the snakes, nearly everything on this island is found nowhere else on the planet. It's primitive here in this land of living fossils; species seem rudimentary, as if the evolutionary incubator has been set to a more languorous clock.

But there is a paradox to all this antiquity. Madagascar, tethered to the planet's distant past, is also an eerie foreshadowing of its future. Madagascar is one of the world's top extinction hotspots. It has among the most unique species to lose, yet nature is deteriorating faster here than almost anywhere else on earth. Madagascar shows what could happen to the rest of the world if today's pace of extinction continues.

That's why I've come here to Antananarivo, the capital city, ten time zones from where I kissed my husband and two children goodbye yesterday in North America. If this is the planet's future, I want to see it now. I've just turned thirty-nine, I'm fighting the knowledge that my marriage of nearly two decades is in its death throes, and I have to see if I can face up to the worst the planet has to offer. And maybe even find redemption.

I arrive with my backpack slung over my shoulder just before midnight. The heat is still so intense that stepping off the plane is like being wrapped in a wet thermal blanket. The sweat runs off me, and my jeans stick to the back of my legs. Madagascar is old-fashioned in ways beyond its odd evolution. There is little industry, little work, few roads and hardly any market economy to speak of. What little economy there is has been in a downward spiral for twenty-five years. Foreign aid is the biggest pool of money going. The vast majority of the Malagasy (pronounced Malgache) live off the land and use wood as their only source of energy, the same as their ancestors have for hundreds of years. In fact, one of the few things growing robustly here is the number of Malagasy who depend on the land for food. The population is growing at nearly 3 per cent a year. This is one of the fastest growing and poorest places in the world.

I've spent most of the plane ride reading a library copy of Richard Leakey's book *The Sixth Extinction: Patterns of Life and the Future of Humankind*. Leakey, the eminent Kenyan palaeoanthropologist and authority on human evolution, is convinced that humans are poised to become 'the greatest catastrophic agent' the world has ever seen, a highly intelligent, highly lethal species set to destroy billions of years of evolutionary advances. Cascades of extinction on this scale have happened only five other times in the history of the planet, most recently when the dinosaurs vanished. Now, he says, we are orchestrating the sixth, a die-off of thousands of critical species, including, possibly, our own.

It's been just over a year since I first learned that many scientists believe that the pace of extinction on the planet has become dangerously volatile, and I've been on a reading spree on the topic. I'm still not sure. What if Leakey and the others who decode the secrets of the bone beds are wrong? Or what if they're on the fringes of scientific credibility? What if more conventional scientists think all this is bunkum? I'm a journalist at *The Globe and Mail*, Canada's national newspaper. I'm not being paid to chase fiction. Plus,

I'm not an expert in biology. In fact, I was a Latin and English scholar, more up on the *Aeneid* than on island biogeography. Much as I've read and interviewed, how can I judge the pace of extinction? And if this theory of the sixth extinction is true, how can I explain something so abstract in ways that people can understand?

As luck would have it, I've arrived in Madagascar with two of the top conservation biologists in the world: Francisco Dallmeier and Alfonso Alonso, both of the Smithsonian Institution in Washington. They spend most of their time in exotic places, looking at the rarest species the planet has to offer and figuring out how to make sure these species survive. Like me, they've come to Madagascar to see what's left of the land known as the last living Eden. But the three of us have also come to look at a richly controversial mining proposal for the south of Madagascar that some proponents are saying could prove a model to help other endangered parts of the world. They argue that the mine could create wealth for the region and the national government and help persuade the Malagasy to use other sources of fuel than the endangered forests. It runs resolutely against the accepted wisdom that business is only capable of wrecking the environment and harming the people who depend on it.

Whatever the merits of the project, the motives look rotten. It is a mine proposed by London-based Rio

Tinto, one of the world's biggest, richest mining companies, to take a mineral out of a piece of the world's most endangered ecosystem in one of the world's poorest nations. It's a very tough sell to some of the international non-governmental organizations trying to preserve what's left of Madagascar's environment. So tough, in fact, that the project has become a cause célèbre over the past decade for the international group Friends of the Earth, which has picketed Rio Tinto's annual meeting in a bid to stop the mine. The anti-mine campaign picked up steam in 1994 when the British environmentalist Andrew Lees of Friends of the Earth died on New Year's Eve right here in Madagascar on one of the proposed Rio Tinto mine sites as he was gathering evidence for the campaign. More persuasive still, Sir David Attenborough – famed for his beautiful BBC documentaries on wildlife and the planet's ecosystems – joined the anti-mine campaign as a tribute to Lees. No other industrial proposal in the world is under more international scrutiny.

Flying over Madagascar the next morning from the capital to the island's south, it's easy to see why there's so much concern. The island should be thickly covered with trees. But instead of the living green of vegetation, the land is pitilessly scoured. The rivers, once clear, run deep red with the rootless earth that washes ceaselessly into them. His Royal Highness Prince Philip, former honorary chairman of the World Wildlife Fund, once said that from above Madagascar looks like a giant animal bleeding into the sea. Less than 10 per cent of the forest cover is left. That's about 6 million hectares, broken up into desperate little pockets running the length of this island, some standing only because they're so remote the Malagasy can't get at them to cut them down. This is not from commercial logging. The trees fall at the hands of poverty-stricken Malagasy, who need to feed their children. It's one of the most massive modern ecological disasters yet catalogued, and it has unfolded mostly over the past thirty years.

But that remaining forest cover is not stable. The trees are vanishing at the rate of 200,000 hectares a year as the Malagasy cut them down for fuel or burn them off in a primitive land-clearing practice they call *tavy*. It's one of the least efficient ways in the world to use land, and the Malagasy are expert at it. Slash and burn the same piece of land often enough and it becomes barren. Do it for long enough and forest disintegrates into sand. And what of all the animals, birds, insects and plants that need the forest to survive? Some move on to find another slender remnant of forest. Many die in the blaze. This is why so many of Madagascar's unique species are on the brink of extinction. They've simply got nowhere to live.

The best example of that is the lemur. It's a primitive version of the monkey and exists exclusively here in Madagascar. It's a primate, a cousin to humans, although privately scientists have been known to say that it's the most dim-witted member of the family. Primates as a whole are horrifically endangered. One in three, or 195 species, are right on the edge of extinction, according to the World Conservation Union (IUCN). Some of the species are down to a few dozen individuals left in the world and others have only a few hundred. That doesn't include humans, the only primate whose numbers and reach continue to grow. But of all the primates on death row, some of those closest to the edge are lemurs. One of the main goals of this trip that Dallmeier, Alonso and I share is to see a lemur in the wild while there are still some left to see.

FORT-DAUPHIN, where we're staying, is a seventeenth-century French fort town of forty thousand ringed with white beaches at the very southern tip of the island. It's got a small airport and a few rugged roads and is the closest town to where Rio Tinto's mineral strikes have been found. That means this town will be the most affected if the mines are built. Shoes, uncommon in Antananarivo, are unknown to most of the people wandering these baked streets. They are dressed in faded, unaccountable clothing – one man is wearing a Western winter coat in the blazing sun as if it were a warrior's cape – most sent here by aid organizations from people in richer countries. At first blush, it feels as if this is the dumping ground for the developed world's cast-offs. Everywhere, the smell of burning wood hangs heavy in the air.

Already, Daniel Lambert, the head of the company trying to set up the mine, and his crew are among the biggest money-spinners in town, with their four-by-fours and drivers, expense accounts and nonchalant Western standards. The group of us take over the Miramar Hotel. Johanne Leveillé, a Montrealer who is working with Rio Tinto and is the only other woman on this trip, whips out a metal container of industrial-strength insect repellent that she picked up on the way, in Paris. It's banned in Canada. Grim-faced, she sprays her room, mattress and pillows – and mine as well – for any multi-legged creatures. Dallmeier and Alonso are horrified. They've been enchanted by the bizarre bugs of Madagascar, crowing to each other every time they spot something they've never seen before. They try to explain that insects – even malarial mosquitoes – are part of the joyous web of life and that the chemical in the spray will undoubtedly do me more harm than any living insect.

Next morning, Dallmeier, Alonso and I try to shake off the jet lag with a swim in the chilly Indian Ocean, a scramble down the hill from the Miramar. Dawn is just breaking. The white-sand beach seems to go on for ever. A couple of huts on the hillside are the only evidence of human life. Then the Malagasy fishermen arrive, paddling into shore after the night's fishing.

They're in dugouts carved from centuries-old Vintanona trees, carted here long ago from a mountain rainforest far away. We pass along gentle hellos in their language, our heads just visible above the waterline. The scene is achingly beautiful, the way it must have been thousands of years ago. It feels as though the rhythm of destruction that grips the rest of the island has abated here just now.

But the passing of time has left something of immense modern value on this beach and several others throughout southern Madagascar. It's the jet-black mineral ilmenite casually mixed in with the bleached sand. Rio Tinto has found three deposits covering about 6,000 hectares underneath some of Madagascar's highly endangered littoral – or coastal – forests. It's one of the biggest strikes found so far in the world and makes up about 10 per cent of all the known ilmenite. Together with the other deposits Rio Tinto has control of, this would give the company a hefty share of the world's ilmenite for the next three generations. Lambert, president of the Rio Tinto/Malagasy company that wants to set up the mine, is here on the beach too, fastidiously decked out in the snorkel and mask he always wears in this ocean. He reaches down and carefully brushes some grains of the heavy black mineral off his feet so he can get his water shoes on before he wades in. That's a sign of how easy ilmenite is to mine. All you have to do is suck up the sand and whip it around in a big separator bin with a large volume of water. The light, white sand spins off, leaving the black ilmenite behind. Then you dump the white sand back where it came from and move on to the next batch. The surf on the beach is doing that now, laying down the black ilmenite like a slash through the sun-struck sand.

After it's separated from the sand, the mineral would be shipped away and transformed from black ilmenite into stark white titanium dioxide. Titanium dioxide is one of the most imperceptible minerals in the world. Its main job has been to replace lead in paint, but it's also what makes toothpaste white and sunscreen impenetrable and fish fingers appealing instead of ashen. The thing is that to mine the sand the company would have to cut down every stick of tree, shrub and bush on top of it. And here in Madagascar, not only is every one of those endangered, but they are all that keep people from starvation. These fishermen taking in the night's fish aren't going to think – like Lambert does – about the abstracts of corporate profits or government revenues or long-term regional economic development when they can't make a fire to cook their catch.

I'm still hesitant to bring up Leakey's sixth-extinction theory. Almost embarrassed, in case it's not polite talk in august scientific circles. I mention it to Dallmeier, who is now standing up to his waist in the ocean. To my surprise, he takes it very seriously indeed.

What's happening now is clearly cause for alarm, he says. I mull over what I have read. The extinction rate is anywhere from one thousand times to ten thousand times normal. And it could become dramatically worse very quickly because so many more species, like the ones here in Madagascar, are right on the edge of vanishing from the planet's genetic store. This is a sign that the structure of life is poised to collapse catastrophically, the way it has done just five times before. The list of these previous five mass extinctions is well known from the fossil record: the Ordovician crash 440 million years ago; the Devonian 370 million years ago; the Permian 250 million years ago; the Triassic 210 million years ago; and the Cretaceous 65 million years ago, when the dinosaurs died out.

This time, though, it's happening because of what humans are doing. But we keep on as if we don't understand that we too are at risk. Why is it that we are so fixated on keeping the death of individual *Homo sapiens* at bay – with vaccination programmes and anti-cancer research and intricate cardiac operations and famine relief – but we don't spend nearly as much time and money making sure the species as a whole can survive?

I stand in the ocean, surrounded by preternatural beauty, and ask Dallmeier this question: are humans a suicidal species? Dallmeier crosses his arms over his bare chest and weighs his answer. He is a careful scientist in charge of an international-class programme at a sober institution. Finally, he says that the prospects for human life are unknown; in fact, they are uncertain.

IT IS MIDDAY and we are buckling ourselves into neon orange life jackets under the cauterizing sun. I've never before been south of Europe, and here in Fort-Dauphin we are not only below the equator, but also below the Tropic of Capricorn. I can see my exposed skin turn pink under the high sun. The sweat and humidity have already proven too much for the only pair of jeans I've brought. One sharp knee bend and they split across the back of the thigh. We're going by motorboat to Evatraha, near the geographic centre of the ilmenite strikes and one of the three places the mining company may use to build the deep-sea harbour it needs to get the minerals out of Madagascar. Local Malagasy, trying to catch enough shrimp in the brackish water so they can feed their families today, stand barefoot on the shore of Lake Lanirano as we pass by. Some are fishing from dugouts. They peer at us. The motorboat in itself is a spectacle. But the stifling life jackets make us look as though we're from another galaxy altogether.

Deeper we go into the secret passages that connect this network of waterways. It feels as though time is collapsing and we have returned to long before this strange island split off from the mainland of Africa 160 million years ago and long before it set out on its own strange course of evolution. Surrounding us in this steamy marsh are bizarre, palm-like plants known here as elephant ears and travel palms, thick as a man's waist at the base leading to slender fronds on top. I doubt I'd be surprised if a dinosaur lumbered up and took a leisurely bite off one of them. The Malagasy boatmen who are steering us keep stopping, backing up, shouting directions at each other and trying new routes to get through. Like so much in this imponderable country, the way forward is not clear.

Finally, the boat eases into Lake Ambavarano and we can again see the open waters of the Indian Ocean. Already, many of the five hundred or so villagers of Evatraha are gathering on the shores, drawn by the unusual sound of the motor and the visiting foreigners, or *Vasahy* (pronounced Vasah). The place is full of children. Many are naked, their bellies bloated by starvation and disease. Some have blond streaks in their curly black hair, a sign of malnutrition. The jumble-sale clothes from the West have penetrated even this far, except that now they're in even worse shape, just shreds grown colourless from the scorching sun. We've been downing bottled water all the way here to beat off sunstroke. Cholera has already killed about fifteen thousand across the country this season, and we're taking no chances on drinking local water. My companions, more attuned than I am to poverty this deep, have left their plastic bottles in the boats, but I take mine into the village. The children clamour for the empty bottle as I drink the last drops. One of their mothers finally scores it, a Westerner's piece of plastic rubbish that instantly becomes a treasured possession in one of the most remote villages in the world.

The village huts are laid out in neat rows spreading back from the beach, a collection of shaky wooden structures standing on sand. Their roofs, woven from travel palm leaves, rise in peaks. The largest have an area not much bigger than a campervan and might sleep a family of ten. The village has a single toilet and it's up a hill, out of the way. Toilet may be too grand a description. It's a flat concrete circle with a hole in it, laid over a pit, lugged up here at the cost of unimaginable sweat by some workers from Azafady, a British aid organization. It's the best the West can offer to help combat the serious cholera outbreak. Short walls don't quite surround it. One of the elegant Malagasy men who now works for Rio Tinto stands guard for me. '*C'est pénible*, eh?' he says, blushing slightly.

But even a toilet this primitive is unsuitable for the people of Evatraha to use. Their legends tell them it is wrong to defecate more than once in the same place. So they do it all around the village, except in the sacred forest nearby. It's an old tradition that they have no intention of changing, even though the Vasahy try to convince them to. Another heavy cement ring stands against a shed nearby, a second toilet that never got installed. The workers from Azafady wanted to put it in the village proper, but the villagers wouldn't let them. They didn't see the point.

It takes me a while to understand what's not here. This should feel like a cool and shady tropical forest. Everything should be green and alive. Instead, it's a desert. There's only a fringe of trees left near the village, right along the bay where we docked, a narrow ribbon of shade against this ceaseless sun. The hills are barren. Rocks show through the soil, the result of the slash-and-burn *tavy* practice performed over and over. The midday heat is unremitting.

The villagers notice that the hills do not grow trees any more, that the crops of rice and manioc are poor. They know that the tasty lemurs they used to hunt with ease around here for their stews are harder to find now. They understand that they need to walk further and further every day to cut the wood to cook that day's food.

Clément Sambo has talked with the people of Evatraha about the changes. He's one of Madagascar's eminent anthropologists, now working with Rio Tinto examining what social effects the mines might have. He points to the square-shouldered village matriarch and the men sitting impassively in the band of shade at the shoreline, hacking the tops off some green coconuts to get at the juice. Even though they know the trees no longer stand around the village, it is impossible to explain to them that they are destroying the trees, he says. He explains that showing them a series of aerial pictures of the retreat of the trees over time would mean nothing. It's far too abstract for people who live day to day like this. They just laugh. How could they hurt the trees? The trees carry their own magic of regeneration. Why should they worry? If they keep walking, they tell Sambo, they will always find another tree.

Evatraha at high noon is only the beginning of the gruelling day Lambert has put together for us. Dutifully, I continue to drink as much water as I can as we transfer from motorboat to four-wheel drive for a jolting hour-long trip to the heart of Rio Tinto's case to the international community that the ilmenite mine should be built. It is *la pépinière*, a tree nursery in the protected forest of Mandena on the edge of a prospective mine site, run by Manon Vincelette, a Canadian forestry engineer. Her task is to recreate the mysteries of this highly endangered littoral, or coastal, forest and help it spread through some conservation areas near the mine site.

As well, she will be in charge of replanting some of the mine site with trees that are alien to Madagascar – eucalyptus and acacia – because they grow faster than the native hardwood trees the Malagasy love to burn for fuel. Already, the Malagasy in other parts of the country, including around the capital of Antananarivo, are forced to use these foreign trees because there's no hardwood left.

The restoration of the coastal forest is a phenomenally difficult task. For one thing, few of Madagascar's living creatures have been studied. The lemur is the big exception. Vincelette and her crew keep stumbling over species that have never been seen before and that may be unique to Mandena, let alone Madagascar, let alone the planet. No one really knows what's here. Even if scientists did understand what was here, they might not be able to figure out how to put the pieces back together. This is one of the most complex ecosystems on the planet, made up of thousands upon thousands of species that are intricately connected in ways no one really understands. Mine or no mine, protected or not, it is disappearing at a steady pace as the Malagasy cut it down tree by tree. This is not just a test of will or faith. If Vincelette and Rio Tinto fail to recreate a reasonable facsimile of this forest over the coming decade, there will be nothing left. The forest will be gone.

This is not the argument some of the non-governmental organizations want to hear. Friends of the Earth, for example, has been pushing for an ecotourism industry to be established here instead of the mine. They believe that will let what's here stay here without industrial use of the endangered landscape, and bring in enough money to boost the living standards of the locals.

To Rio Tinto, the solution is not that simple. Yes, the mine would strip some of the remaining endangered forest off the Malagasy land, but at least if Rio Tinto were involved, some of that would be recreated, however experimentally. As it is now, the forest is simply disappearing. It's not only endangered, meaning rare on the face of the planet, it is also horribly damaged. Two years ago, there were 589 hectares of forest on the 2,334-hectare site where Rio Tinto wanted to mine first. Of that, 44 per cent was highly or extremely degraded. The rest was just degraded. None of it was healthy. Today, it's much worse. Over just the past two years, the forest has disappeared one piece at a time, carried home over the shoulders of skinny men swinging axes at their sides.

The local World Wildlife Fund office has estimated that about 250 manloads of wood get carried to Fort-Dauphin every day. Even if environmentalists and sociologists and economists could figure out how to reduce that by half today, the population is growing so fast that in forty years it would add up to the same thing again. And it's the same situation the island over. If this keeps up, Madagascar will look like a moonscape, the lemurs will die off and so will the people. And that's despite the roughly $300 million (US) sent to Madagascar from foreign countries over the past decade to save the environment. Things just keep getting more desperate. In a way, Vincelette's *pépinière* is not just the seedbed of Mandena's hopes, but also of Madagascar's.

Rio Tinto stepped into the story as an unlikely hero, a sort of reformed rake with a history – by its own admission – of damaging the land with mines in other eras, in other parts of the world. The negotiations have now reached a critical point. Lambert has been to Madagascar perhaps four dozen times so far as he has steered the mine through talks with the Malagasy government. Now, the company is preparing an environmental assessment on the project that is needed before the Malagasy government will give the mine final approval. The document is bound to draw intense scrutiny from conservation groups all over the world and must break new ground in describing the environmental and social benefits the project could have and justifying the mine.

Rio Tinto is nervous because this is one of the most controversial projects it has ever encountered and because it could raise the bar for what the company will be expected to do in other projects elsewhere in the world. The Malagasy government is nervous because it badly wants its share of revenue from the mine but it doesn't want to look like a villain on the world stage by further endangering a remnant of its fragile paradise. This is a game that must be played with great care and diplomacy.

Dallmeier is taking it all in, wandering around this

nursery at Mandena. He knows that Rio Tinto's mine has the potential to pour much more money and work into Madagascar than any aid organization. Already, Rio Tinto has gone to far greater scientific lengths to protect the environment here than most companies have done anywhere else in the world, even before receiving approval to build the mine. World-class scientists have flocked here over the past few years to do research.

Dallmeier knows that a publicly traded multinational such as Rio Tinto will have to answer to shareholders if it bungles an important project like this, and that it has the financial muscle not only to do the research but also to use it by hiring the best science in the world to make an honest go of the ecological restoration. Britain's Kew Gardens, keeper of the planet's botanical patrimony, has already collected samples of the seeds Vincelette is trying to grow. In itself, that leaves room for hope that restoration may be possible one day if not now, he says.

There's pressure now from the world's financial institutions too. They have become so worried about backlash from the public over ecological destruction that they're making money cheaper for companies that take pains to leave the land and the local population in better shape than they found them. It's an emerging business trend Dallmeier has witnessed in other parts of the world as companies are forced to look for business in more and more remote spots. Dallmeier knows the Rio Tinto project is a tough sell, an uncertain experiment. But he also knows that the situation in Madagascar is so perilous that it may be time to look for novel solutions. The old ones that eschewed the involvement of corporations haven't worked. If this one does, it could show other countries a way out of the cycle of degradation. Doing nothing is no longer an option.

He's under no illusions about how difficult it will be. 'Basically, they have to become greenhouse gardeners with a bunch of seeds and plants nobody has a clue how to grow,' he tells me. As Vincelette, all legs and freckles, explains to us how she set up her *pépinière* with some of the best scientific expertise from around the world, it seems to me that the task to recreate this riotous ecosystem artificially is too big. I fear that it is fated to fail. Humans are just not that smart.

When she came here to Mandena to set up her research station and nursery, Vincelette couldn't even find a potting soil that would grow her seeds. She had to experiment with different combinations of growing mediums. Finally, she settled on a combination of sand, slow-release fertilizer and manure from Madagascar's zébu, a hump-backed cow. She points proudly to three neat piles of this home-made potting soil on the cement floor of the hut where the seeds are stored. It took a year to figure out that mix.

Getting the seeds is another labour of love.

Vincelette has a staff of seed-catchers who go out each morning armed with ragged plastic shopping bags and sticks 10 metres high. Each stick has a fork on the end for prising the fruit or seed casings from the tops of trees. Once the men have captured the fruit or the casing, they plop it into the bag and bring it back here to the hut. They spend the afternoons in a gentle ritual: press the seeds away from the flesh of the fruit or out of the tough casings, taking care not to crush or crack them. Then pan them in water, as if for gold, separating them from every fibre that surrounds them. The clean seeds then rest here on net hammocks for about a month until they are profoundly dry, labelled with little signs lettered in French, the accents precisely marked. The seeds are shiny and dark and queerly shaped. Some are the size of a fat brown thumb.

Outside, among a scattering of sheds built of rough-hewn wood, Vincelette has set up carefully marked frames filled with saplings. They are thriving. Blessed by the tropical heat and long days of sun, some of them grow several metres in a month. Vincelette has done a few trials for large plantations, but that's still highly experimental. Once the mining operation starts, Vincelette's task will be to plant 150 hectares of trees a year to jump-start the ecological processes needed to make this system healthy again. The idea is that once the trees come back, so will other species of plants, animals and microbes needed to form the forest again.

Most of those planted hectares will be filled with eucalyptus. The goal is to persuade the local Malagasy to cut down the weedy eucalyptus for fuel instead of the rare trees that are vanishing so fast. But, like trying to persuade the villagers of Evatraha to use a toilet, this task has no guaranteed success. It's not clear to anyone if the local Malagasy can be persuaded to switch to eucalyptus while native hardwood still stands. In fact, the Malagasy word for tree translates into French and English as 'wood'. Not a living thing, but a dead one.

A sign of how fixed the local legends are may be found in Vincelette's experiment with mahampy, a marsh reed that locals around Mandena use to weave baskets and mats. Like the villagers of Evatraha, the people here don't believe they have any control over the trees. Rice, manioc, sisal: those can be grown, they say. Trees and reeds have to grow on their own. They are at the mercy of the gods of nature. Vincelette finally had some of her work crew dig out a small piece of land to below the water table and plant mahampy. At each stage, she brought in some of the opinion-making women from the villages around Mandena and showed them what she was doing. Two days ago, the women came back and found that the mahampy was growing. Now they've started calling Vincelette 'the second god', a good-humoured poke at her powers and their own changing beliefs.

As dark falls, the smell of wood smoke becomes

thicker, even here in Mandena. I look for the source. Off to one side of the nursery, behind the trays of laboriously planted saplings, the crew's cooks are making a supper of fried fish and rice over wood fires. Smoke billows out of the lean-to. My eyes water. Wood is the only source of fuel for hundreds of kilometres in any direction. To the Malagasy cooking dinner, this is what wood is for, whether the forest is endangered or not.

It's late now and the sky is pitch black. The stars, unrivalled by the lights of human development, seem to press closer to the ground here than in North America. For the first time, I understand why the ancients believed the stars had tales to tell them. Despite the hour, our day is far from done. We haven't seen any lemurs yet in our travels, and Lambert has promised Dallmeier and Alonso that we can search the protected forest of Mandena for some of the ones that come out only at night. This forest should be a safe bet. We strap lights on our heads and set out.

Dallmeier and Alonso have a distinct spring in their step. We cross a bridge and they sigh in delight, shining their beams into the water. Some rare Nile crocodiles. My stomach's doing somersaults. I am on the other side of the planet from my home, in a tropical forest on a strange island in the black of night, and I have no idea where I fit in on the food chain. Again, I have the odd sensation that time is collapsing. It feels as though the forest is painted with all the ancient colours of human fear, honed through the precision of evolution over the millennia to keep us wary in places such as this. No wonder most of the scary bits in fairy tales take place in a demonic wood.

A few metres ahead, Alonso chortles and emerges from a thicket of trees. He has found a type of snake he has never heard about before and is holding it up in triumph while Dallmeier, who is a *National Geographic*-calibre photographer, hurries to take pictures. The others are shining their lights into the trees, trying to catch the glitter of a lemur's eyes. Nothing. They thrust deeper and deeper into woods. Lambert shouts that he's spotted one and some of the group rush after him, hooting with primeval glee. Nope. False alarm.

Dallmeier, meanwhile, is using his light to search for something else: the mark of man. It is everywhere. It's even worse than Dallmeier has expected. Machete marks there, scoring an ancient tree. A massive stump over there where a woodsman has chopped down a tree, protected woods or not. And an unnatural silence. A healthy forest would be loaded with night-time sounds: insects, moths, birds, the stealthy movements of predators. This one, by contrast, seems barely alive.

Finally we make the trek back to the four-by-fours and tumble in. We're streaked with dirt and sweat. I can smell the acrid scent of fear. No lemurs. We have just started the hour-long drive to the Miramar in Fort-

Dauphin when we come upon a lone man, sneaking out of the protected area. He is listing under the weight of a massive old-growth tree balanced on his shoulder. He has spent hours cutting it down on the sly and squaring it off by hand with his axe. Now he is carrying it by foot back to Fort-Dauphin. It's worth maybe 4,000 Malagasy francs, less than $1 (US). That will buy him a few precious handfuls of rice or a little manioc, not enough calories to replace what it took him to get the tree to market. This is the cruel maths of Madagascar.

The next morning, we are determined to see a lemur in a sliver of Malagasy nature that has not yet been razed. Vincelette, Dallmeier, Alonso and I head for Berenty with two top-notch Malagasy scientists, Jean-Baptiste Ramanamanjato and Laurent Randrihasipara. Berenty is a private reserve of 260 hectares and one of the very few pieces of gallery forest left in Madagascar. The lemurs are especially fond of gallery forests because the trees meet at the top, forming a complex series of arches that make a perfect canopy for the creatures to hop from tree to tree. About 150 ring-tailed lemurs live here, next to a research station plus visitors' centre. If we can't find lemurs in Berenty, it's safe to say we can't find any in Madagascar.

Berenty is perhaps 80 kilometres from Fort-Dauphin. But the roads are decayed and the going is torturously slow. The heat is sickening. I reach out and touch the chrome on the door of the four-by-four. My fingers burn. Ramanamanjato is one of the few people in the world who can identify whatever is to be found in Madagascar and, to boot, he has quick eyes. Up there, he murmurs, pointing as we drive past. A Newton's falcon sitting on a sisal flower. Only in Madagascar. Dallmeier and Alonso insist that we back up to take a closer look. A few more kilometres down the road Dallmeier shouts for us to stop again. He jumps out, hops down into a culvert by the side of the road and starts snapping pictures of a spider he's never seen before. Even Ramanamanjato doesn't know what kind it is. It occurs to me that this may be the first time in history that a scientist has seen this particular species of spider. It's possible there are only a dozen left on earth. That's just how things are in Madagascar.

None of us mentions the mountains, but we all examine them silently. Like the hills of Evatraha, they are scarred with slash-and-burn *tavy*. Dallmeier and Alonso have been reading the latest issue of the scientific journal *Nature*. It contains an article listing the twenty-five terrestrial areas conservationists believe have the most unique species to lose, the worst ecological problems and the least protection from further destruction. The authors, including Oxford University's Norman Myers and Conservation International's Russell Mittermeier, call these 'hotspots' because they contain the remaining liveable land of 44 per cent of the world's plant species (133,149 species) and 35 per cent (9,645 species) of the world's vertebrates. Once, these plants and animals had a total of 11.8 per cent of the planet's land surface to live on. Now, that's shrunk to just 1.4 per cent, the result of cutting and fire. The twenty-five hotspots include such places as the tropical Andes, the Caribbean, central Chile, the western African forests and the Philippines. They are the most vulnerable parts of the planet. Of all of them, though, the *Nature* article pegs Madagascar as the most vulnerable.

We pass a woman on the side of the road selling charcoal, or *charbon* as it's known here in French. Heaps of burlap bags of charcoal are piled behind her. She is the richest woman I've seen so far in this country – a new denim dress, gold hoops in her ears and a pierced nose. Locals make the charcoal at the edges of forests, cutting trees down, covering them and burning them. It saves so much carrying. And it means they can cut trees down faster. The bags sell for less than $2 (US) each. Vincelette narrows her eyes. Each one of those bags contains the equivalent of three trees, she says. Maybe as much as five.

'That's where the forest is going,' she tells me.

Finally we arrive at Berenty, a slim island of nature holding out against the surrounding fields of a sisal farm. We drive inside the gates. Boom, there are the lemurs, snoozing in the shade of the gallery forest, out of the blast of the midday sun. A ring-tail rouses itself, eyes us quizzically from its tree. Then it leaps over to take a closer look, as if it has finally found us instead of the other way around. Its long black-and-white striped tail curves behind, balancing. It edges closer and closer, lazy in the heat, but curious all the same. I can't make out whether its sounds are more like chirps or grunts. Maybe halfway in between.

It looks chinless for a primate. Its proportions seem out of whack: the limbs are much longer that I expected. It steadies itself with its hands and then daintily lifts one up and darts its diamond-shaped pink tongue out to lick each finger in turn. Like us, it has an opposable thumb, long, as if it has four joints instead of three. Three fat white sifakas, a lemur cousin also found exclusively in Madagascar, are asleep nearby, draped over the branches of the trees.

Dallmeier and Alonso are enthralled. And not just with the lemurs, but with the forest too. Finally, they have found a slice of intact forest. They know that they are some of the few people on earth able to see a forest like this today and that the chance to do this may soon be lost for ever. The two of them breathe it in, eyes darting everywhere, on the lookout for every species they've ever read about. It is unfathomable to me that so many hundreds of species of trees can live together in a single ecosystem. This is not what I've seen in North America or Europe, where a forest is made up of three or four types of trees, rather tidily laid down. Here, chaos has taken over. Everything seems intertwined, interdependent, riotous. It is verdant and unfathomable and fully alive. Finally I understand. This tiny little piece of gallery forest, standing in the middle of a wrecked land, shows me how the planet itself works. Take one piece out of this and things will probably be OK. Take enough, though, and nature will not answer for the consequences. Eventually, places like Berenty will live on only in memory or, perhaps, as a dream.

As I think about it further, I realize that even if Berenty were left alone it could still not carry on for ever. It too has a shelf life. It is a small slice of forest, isolated from everywhere else on the planet. The lemur gene pool here in this island of life has got to be getting thin. And there's no obvious way to replenish it. Over time, that means the lemurs here will weaken genetically. One batch of illness or a bout of rotten weather could take them off. Over time, they are fated to die out. In Madagascar, it's gone far past the imperative to do no more harm. Now, the task is to restore. If that's not clear here in this ecological disaster zone, where will it be? And is our lethal, naive species smart enough to know how to restore, even if we recognize the need?

It strikes me that the Victorian gentleman-naturalist Charles Darwin is a character in this story too. I can't help thinking about the remarkable voyage he took around the world in the 1830s aboard the *Beagle*. He passed within spitting distance of Evatraha on the morning of 31 May 1836: 'In the early part of the passage we passed in sight of the south end of Madagascar . . .' he wrote in his journal. That trip, launched just after Christmas in 1831, led Darwin to understand and eventually explain how nature works. His findings horrified him. He refused to publish them for a generation, unwilling to undermine the theological beliefs of the day that told people a different story of how life came to be.

But if it was Darwin who first glimpsed how species reacted to a changing environment, then surely his journey – on the *Beagle* and afterwards – must contain some clues for a society like ours, teetering on the brink of a mass extinction that might shorten our own shelf life. A reluctant revolutionary, he managed to convince his society to examine the legend that ruled the understanding of the time and then wholly rewrite it.

He had no way then of knowing about the five big extinctions or that a sixth cascade of extinction would loom just 150 years after he first published his theories. But faced with the evidence before us today, would Darwin have anything to say to us? If his theories could explain how past species came to be, must they not also explain how future species will evolve and therefore what the risks are today?

Again, I think of the Milman quote: 'History to be true must condescend to speak the language of legend.'

It strikes me that this is the key to understanding why we're not doing something about the ecological crises of the planet when we could.

It's about the legends of society. Some are universal. That means they are 'so deeply embedded within cultural consciousness that much subsequent narrative is infused with their imageries', as Penelope Lively writes in her introduction to *The Mythical Quest: In Search of Adventure, Romance & Enlightenment*. Among these would be the grand quests for knowledge and truth, the fundamental battle between good and evil, the fall before the rise. And on a human scale, this would be the metaphor for the difficult journey through life with its disappointments, its triumphs, its acknowledgements of human frailties and insecurities.

Other legends are very, very local. I think about the villagers near Mandena who don't understand that they can grow mahampy. The villagers of Evatraha who don't defecate in the same spot twice, who don't understand the link between using toilets and preventing cholera. Who don't think they can hurt the trees. There's no sense explaining to them the dangers of *tavy*, or trying to convince them to switch to eucalyptus or solar panels until they understand how biology works. Their legend tells them that trees cannot disappear, despite the evidence of their own eyes, which no longer see trees, or their legs, which have to keep walking further and further to find them. Until they can write another legend to replace that one, the terrible dilemmas of Madagascar cannot be solved.

People in the West have long since put aside the legends that inform these Malagasy beliefs. It would be easy to chuckle at the Malagasy and their inability to understand the abstract thought required to draw a line between planting mahampy and growing it, between toilets and prevention of disease, between cutting down too many trees and cutting off the land's ability to renew them quickly. It would be easy to scorn the Victorians who believed that no species had ever gone extinct and none ever would because each one was set on earth by God.

But how different is that from the modern inability to accept our own vulnerability and that of other species and then take pains to preserve our planet's life-support systems? Is that not the same quality of abstract thought required of the Malagasy, the Victorians? Is it not the same challenge to legend?

In Madagascar, it's easy to see the destruction, the retreat of the trees. It's harder to see in other parts of the world, but a vast array of scientific research tells us that ecological crises are urgent and worsening. Granted, the solutions to all these problems are unknown. But until we accept that the problems are there, that the planet's resources are exhaustible and we are trying them sorely, we won't really look for an answer. We will continue to believe that if we keep walking, we will find another tree.

THREE

Reading the Secrets OF THE Fossils

Time does not easily give up her secrets. But here in the fertile plains of North America she has laid bare a select few.

These are the famed bone beds of southern Alberta, where literally millions of fossilized dinosaur bones are exposed in a queer little swathe of landscape that cuts through the middle of the continent. But as rare as it is to be allowed to peek into the records of time, this particular slice of the planet has something even rarer, even more riddling to reveal. It is the only place discovered so far on earth where it is possible to see what was happening on land – as opposed to in the sea – when a massive asteroid hit the planet 65 million years ago.

Scientists can tell from what they see here and from

the marine fossil records from that time that it was a cataclysm, that perhaps half the earth's life forms did not survive, that the planet itself suffered a shock. That moment, making up a layer perhaps as much as two centimetres deep within the fossil record, marks the end of the Cretaceous period and the beginning of the Tertiary. This sliver of history – representing perhaps a decade – is known as the K/T boundary (for Cretaceous/Tertiary). It is the incredibly rare physical marker of a single point in time in the planet's 4.6-billion-year history.

Of all of the secrets of the past, none captures the imagination of palaeontologists more than this one. Fossil diggers have found the remains of dinosaurs from 150 million years before the K/T boundary, and it's clear that the creatures were so varied and so numerous they ruled the earth. But after the boundary, dinosaurs are absent. The rulers of the planet went extinct in the geological equivalent of a heartbeat.

It's widely accepted that the asteroid killed off the dinosaurs. There are reams of information on the big hit. Geologists can now tell us when and where the asteroid came down and what happened to the planet as a result.

The geophysicist Walter Alvarez, for example, says it was the equivalent of a hundred million hydrogen bombs going off. Rocks vaporized as the asteroid struck, rising in a suffocating cloud back into outer space.

Animals burned instantly to a crisp, fried by the very surface of the super-heated planet. Tidal waves caused wanton destruction. Fires engulfed whole continents, filling the skies with soot. Then the darkness came. And the cold. Months later, the carbon dioxide that had been loosed from the earth's surface by the explosions began to heat the climate to lethal temperatures: a greenhouse effect gone mad. Centuries later the carbon dioxide pollution seeped away, until eventually the temperatures returned to normal and life could thrive again, minus the dinosaurs that had been the mightiest creatures on the planet. It was a geological day of reckoning, an evolutionary cataclysm.

But palaeontologists aren't so sure. Yes, they say, it's clear that an asteroid hit the planet. But did it wipe out the dinosaurs? That's a whole other story.

Philip Currie, who is one of the most famous palaeontologists in the world, has been thinking about this for years, almost by default. He's made his reputation finding out about the dinosaurs when they were in their prime rather than searching for answers to the great mystery of why they vanished. But he is based here at the Royal Tyrrell Museum in Alberta on top of the best evidence in the world on the K/T boundary, and he just can't stop himself from digging into this puzzle too.

The bone beds have told him a tantalizing story, which he recounts to me one evening over dinner in

Calgary. I've been telling him about Madagascar, what I've been finding out about the sixth extinction and my fears that we humans are shortening our tenure on earth.

He jolts to attention. He is trained in scientific dispassion, but this topic cuts a bit too close to home. He too has been conscious of the sixth extinction currently under way and the threat to humans, he tells me. But he's been looking at it by examining the fifth extinction, when the dinosaurs died off. He's not sure yet whether his theory holds up but, if it does, it has terrifying lessons for *Homo sapiens* at just this point in our biological reign on earth.

The theory is this: the dinosaurs, which were the most complex, highly advanced animals to have lived on the planet until then, were struggling with the breakdown of their ecosystems millions of years before the asteroid hit. Some species had already lost the struggle and gone extinct. The fury of the asteroid was the final, fatal blow. That is, if there were any dinosaurs left at all by the time it arrived.

Currie tells me that if the asteroid had hit, say, 10 million years earlier, when the different species of dinosaurs were still numerous and their gene pool much deeper, some of the creatures might have survived. The Age of Mammals, which got going only after the stronger, wilier dinosaurs vanished, might never have happened. The Age of Man would be science fiction. Evolution might have taken a radically different course.

Come to the bone beds, he tells me. Travel from Madagascar, land of the living fossils, to the North American plains, land of the painstakingly excavated ones. Look at the sixth extinction through the prism of the fifth.

I ARRIVE AT HIS MUSEUM in early spring. This is the westernmost edge of the North American prairie, just this side of the Rocky Mountain range that forms the ragged backbone of the continent. To me, it has always been the land of light. Sometimes the sun's light is so sharp that it bleaches definitions away. At other times, it puts things into sharper relief, gives them more definition. But whether the sun masks or unmasks, on the prairie it always finds a way to teach me more about what place humans occupy in the universe.

Today, the prairie is bitten by a keening wind. There was a freak blizzard yesterday, and a layer of snow remains on the ground. Currie shivers in a light jacket. His hands are in his jeans pockets, his eyes to the ground. Like many bone hunters, he has unusually long legs, a strategic advantage in a profession that requires you to scramble over unforgiving terrain. His face seems to relax only when he is squinting into the sun.

He stoops. Pokes a long forefinger into the dust on

the ground and uncovers something. It looks like any old piece of sandy rock.

It's maybe a couple of centimetres in diameter and about the same in thickness. One side is flattish and fairly smooth. The rest is rough to the touch, pocked with what look like irregular pinpricks.

This is a piece of fossilized dinosaur bone, likely about 70 million years old. He can tell that much just from its texture, colour and where he found it.

But he can also make this dusty bone come back to life: this was a plant-eater.

When it was alive, that roughness was the spongy inside of a bone. Spongy material like this evolved over hundreds of millions of years to reinforce plant-eaters' bones so that they could grow bigger and heavier. He figures this one was a duck-billed dinosaur called a hadrosaurus.

The bones of meat-eaters, like the *Tyrannosaurus rex*, one of his specialties, took another course of evolution. They were hollow but extremely dense and strong on the outside. That density is what carried their tons. No sponge.

Moments later, he crouches. His long legs fold up like a cricket's. Triumphantly, he prises loose something else that looks just like a square bit of rock. This one is perhaps a centimetre square and a quarter of a centimetre in depth. He squints at it. 'Tendon!' he announces.

That means this was a duck-bill for sure. They had tendons running down the spine to stiffen the tail, he explains, holding the fragment of fossil in his palm.

He makes it seem so easy. But to be a palaeontologist is to light candles every day to the gods of luck. The fact that the fossils are here in this spot to be found at all – in these tiny but important shards as well as in full skeletons – is sheer luck, the happy confluence of three factors.

Currie summons up the past to explain what they are. The hard-pan prairie we are standing on disappears and we can smell the magic of this place. We have gone back in time by about 80 million years, to when the dinosaurs of North America were plentiful and strong. Most were the size of deer or humans. Only a few were the behemoths we now associate with dinosaurs. This part of the planet was a coastal lowland on the edge of a massive sea. The Gulf of Mexico ran to the Arctic Ocean and the polar ice caps weren't around. The climate was sultry, much as it is today in Florida. Instead of the stern scrub that dots the ground today, this part of the world was lush with forests of cypress and redwood, lily pads, ferns and some of the first plants to have flowers. The air was heavy with the scent of magnolia.

The first factor that made this part of North America a future palaeontologist's dream was the vegetation and the even climate. That meant that herbivore dinosaurs gathered here, attracted by all the plants, and carnivore dinosaurs followed.

Second factor: because these dinosaurs lived so close to the water's edge, the chances were good that when they died their bodies would be covered quickly in sediment. These are the ideal conditions for fossilization. Even so, Currie reckons that only about one in ten thousand dinosaurs remains intact enough to be fossilized. If dinosaurs died in forests, they were likely to be trampled or dismembered or eaten by bacteria long before they could become fossils. In a lush environment their bones simply disappeared, corroded by the acid of the swamp.

Third factor: tens of millions of years after the dinosaurs lived, died and were preserved, the glaciers came through. That was about twelve thousand years ago, the recent past in geological terms but a time that *Homo sapiens* call prehistory, absorbed as we are in our own short span on earth.

As the glaciers retreated, they carved pitilessly through the landscape. They peeled off several hundred metres of geological beds – or more than 65 million years' worth – including any trace of the Age of Mammals. In doing that, they exposed the geological record of the glory of the dinosaurs. And, just here, a glimpse of the K/T boundary, undiluted with records of marine life. Currie points to a bald outcropping, sharp against the cold, clear sky. And then another. They are rounded, as if a gigantic hand had scooped down and sculpted them.

Currie is the fourth lucky ingredient in this mystery, although that's not the way he would put it. Dinosaurs captured his imagination when, as a child, he shook a plastic one out of a breakfast cereal box. He has been enthralled ever since and has won renown for being able to keep his thinking on dinosaurs fresh and original – out of the box, so to speak. He was an early proponent of the theory that modern birds are the only descendants of dinosaurs. That's now mainstream.

And he has learned to read these layers. The black seam was a mucky wet time; the grey shows when forests covered the land; the sandy one is the residue of an epoch of fast rivers. And because the layers are so precise, Currie can trace time through their formations. That means he can also tell which dinosaurs lived when. That part he pieces together not just from intact skeletons, but also from the broken and elusive bits of fossil he has found for me today. In fact, some of his most important theories have come from these fragments, less thrilling than the intact skeletons but, in some ways, more revealing.

It is the story of evolution, of a random and remorseless nature, playing out tens of millions of years before Darwin set foot on the Galapagos Islands and started to pull all his heretical pieces together. And it is the story of time. Not told in the impatient human frame of months and decades, but rather in thousands and millions of years.

The surprise ending to the story Currie reads here is that there is no evidence the dinosaurs survived to the K/T boundary. There's no evidence in the fossil record that they suffered an abrupt mass extinction. It's much more likely that they died out slowly over time. There's plenty of evidence that the ecosystems the dinosaurs needed to survive were changing rapidly. Maybe too rapidly for the dinosaurs to keep up. It could be that their time had come before the asteroid hit.

Coming to this conclusion is something like putting together tiny fragments of a large *T. rex* to make a skeleton. A lot of the information is missing, or sketchy, especially on land. The record is much better from the sea, simply because so much of the planet is covered by ocean. And while creatures in the sea could move about freely and therefore be distributed widely, it was a different story for massive land-based dinosaurs that may well have existed in rare pockets across the land.

So far, any evidence about the dinosaurs at the end of the Cretaceous period is from western North America. There are no sites in Europe, Australia or Antarctica. None has been found in Asia or South America. And the word from western North America – meaning here in Alberta and in similar sites in Montana – is that there are no fossils at the boundary. In fact, here in Alberta, there aren't any dinosaur fossils within

10 metres of the boundary. Currie shakes his head as he thinks about it. That could represent an absolutely enormous amount of time. Down in Montana, they've found a dinosaur fossil within 1 metre. But still nothing closer. If the asteroid destroyed all the dinosaurs in a geological instant, wouldn't there be fossils at the boundary?

What the fossil record does show, though, is disturbingly clear. It's as simple as taking a canoe ride down the Red Deer River towards Montana. You travel backwards in time, seeing bone beds from 64 million years ago, then 70 million years ago, and then from 75 to 80 million years ago. That runs from a million years past the K/T boundary and on through the final 10 to 15 million years of the dinosaurs' rule on earth.

In the oldest beds, Currie and other palaeontologists can find thirty-five species of dinosaur. By the time they get to the beds that are 70 million years old, the species list has dropped to between twenty and twenty-five. Near the boundary – but quite a bit before the asteroid hit – there are just six. That's a remarkable loss of diversity within that time frame.

And the six dinosaurs that made it to the very end of the Cretaceous period make up a bit of a motley crew. Four were very highly advanced and specialized, and two were among the most primitive ever found. Not only that, but the final dinosaurs to walk the earth were the biggest: tyrannosaurus and triceratops, for example.

That's a clear sign that the climate and environment had been stable for a long time because evolution favours big over small under those conditions. In fact, big is normal in evolutionary terms. The fossil record shows this over and over, although it seems an odd notion today because animals are relatively small now compared with other times in history. Humans have killed off most of the big ones that we used to share the earth with.

Bigger animals have a longer life span. They are less active. They need to eat less for their volume than small animals. They reproduce more slowly and therefore adapt more slowly to change. Their shelf life is robust as long as climate and environment don't suffer sharp shocks.

Once upon a time, for example, there was a rhino that stood 5.5 metres at the shoulder. There were amphibians with skulls as big as a dining-room table, bisons the size of elephants, land-living birds that stood 3 metres in height. And, of course, the monstrous dinosaurs.

Towards the end of the Cretaceous period, when the huge dinosaurs won the evolution stakes, it looks as if other dinosaurs had already lost, and died out. The gene pool of dinosaurs became thin, Currie believes, and before it could replenish itself, disaster struck, in the form of the asteroid. If there were any dinosaurs left by then.

But this is a mystery wrapped in another mystery, I say. Why would the dinosaurs have been dying out? It was probably ecosystem collapse, caused by the beginning of continental climates, with their warmer days and colder nights, Currie says. It was the Cretaceous version of climate change, a subtle but highly significant change in the earth's average surface temperature. He pulls me back to the present, sends me to a room in the museum where his colleague Dennis Braman works. This part of the story must be told by reading yet tinier particles of the earth's past.

Braman, who specializes in identifying plant microfossils, is holding a piece of the K/T boundary in his hand. It is a sliver of claystone sediment 1 to 2 centimetres thick, packed between layers of mudstone and coal. He has taken this boundary apart spore by spore, molecule by molecule.

It shows a bunch of substances not normally found on this planet. There is an astonishing amount of iridium, one of the six extremely rare platinum-group elements bunched together on the periodic table. The iridium is so abnormally concentrated here that it's become known as the iridium spike. That spike has been found along the K/T boundary all over the world. Extra-terrestrial bodies such as meteorites are rich in iridium.

Then there are all sorts of shocked quartz. That's quartz striated with so many fine fractures running at cross angles to each other that it resembles miniature graph paper. It's a full-scale destruction of the crystal's integrity, known to happen only after a short – but extremely intense – burst of pressure. Like that from an asteroid slamming into the planet.

It's the same story with the loads of unusual primordial microdiamonds, common to extraterrestrial bodies. They suggest that a shower of diamond dust blanketed the planet. Just 2 centimetres on either side of the boundary, the diamonds are absent.

And then there is the array of bizarre amino acids, says Braman. He runs his finger along the slender layer of sediment that marks the boundary as he explains.

Living systems on earth contain fewer than thirty amino acids, the building blocks of proteins. The boundary contains more than fifty, many of which are unknown to occur naturally on the planet. But they are common on asteroids.

So it's clear that something catastrophic happened. Right above the evidence of the asteroid, there's what Braman calls the impoverished zone. It's the right type of sediment to preserve plant spores, but hardly any are there. A few species, but hardly any specimens. Strangely enough, though, there's strong molecular evidence that odd things were going on long before the catastrophe.

The main clue? Before the iridium, microdiamonds, shocked quartz and extraterrestrial amino acids, not only do the dinosaurs disappear, but most of the plants do too. Throughout the Cretaceous fossil record, there's evidence through spores and pollens of a vast, rich array of plants, including a complex upper layer of canopy forest. All of a sudden, most of these are wiped from the record. Instead, the ferns take over. As much as 90 per cent of the plant life was fern during this queer pre-cataclysm time, Braman tells me. That's completely uncharacteristic of the rest of the Cretaceous period.

Ferns are survivors. They take over when other plants die out, sort of the Cretaceous equivalent of fire-weed springing up after a fire has swept through a modern forest. Braman's theory is that, as the other types of plants died, ferns seized the opportunity and took over, spreading to parts of the land where other plants had crowded them out before.

All this tells Braman that the ecosystem that supported plants – and dinosaurs – was unstable well before the asteroid hit. This was a textbook ecosystem collapse. It is Darwin's theory of natural selection played out millions of years ago.

As I puzzle through all these ideas, it strikes me that humans have been trying to read the fossils for thousands of years. And that every story we come up with tells us more about how we see the world than about what time really has to tell. In other words, we see in the fossils what the legends of our day allow us to see. We see what is culturally knowable.

North America's indigenous peoples had a story for the massive fossilized dinosaur bones they found here. They believed they were the bones of giants who had once walked the earth. Ancient Romans and Greeks believed that amber (actually the fossilized resin of ancient trees) was formed from the tears of the sun god's daughters – turned into poplar trees by the shape-shifter Zeus – crying at the death of their brother.

An even more enduring legend comes from the remains of the sophisticated Greek, Roman and Minoan civilizations, long before thoughts of evolution or extinction or dinosaurs had pierced the human imagination. This is the tale of the griffins, the mythological hounds of Zeus said to be born of the mating of an eagle and a lion. They had the body of a lion and an eagle's beaky head. Sometimes they had wings and a mane of spiral curls. They were deadly yet sacred, bred for guarding Zeus's gold.

To the sophisticated thinkers of that civilization just a few thousand years ago – the recent past in geological terms – this made perfect sense. Humans were bit players in the cosmic game. Gods could mix and match any creatures into fantastical combinations at their whim.

Today, some palaeontologists, including the famed American dinosaur hunter John Horner, and some anthropologists theorize that the idea for the griffin came from white fossilized skeletons of the dinosaur protoceratops found millennia ago in the red cliffs of the Gobi Desert. To the eye of a sophisticated Minoan, these fossils look just like a parade of griffins. They were satisfying proof of the stories of the gods that they told themselves.

Attempts to solve the mystery of the fossils didn't end there. Less than two hundred years ago, around the time Darwin was boarding the *Beagle* and figuring out the theory of evolution, the eminent scientists of the day were tying themselves into intellectual knots trying to explain the dinosaur fossils that were emerging from the depths of the ground in Britain and Europe with increasing frequency. They couldn't figure out how these strange old bones fitted in with the Christian maxim of the fixity of the species.

In the past two years, I've seen two of the most beautiful and oldest scientific collections of dinosaurs, in museums in Berlin and Oxford. The University Museum in Oxford, filled with magnificent wooden cases hinged with brass, tells us that in 1677, during the Renaissance in Western civilization, the early bone-finder Dr Robert Plot found a megalosaurus thighbone near Cornwall. It was from the Jurassic period, perhaps 168 million years old, but Plot believed it had to have come from a giant human.

Fossilized iguanadon teeth found in 1822 stymied some of the best scientists of the day, including William Buckland and Georges Cuvier. (Cuvier was later a central figure in the Darwin debates.) Could they be from fish? they asked. Maybe a rhinoceros? It wasn't until 1842, when Darwin had been back from his travels for six years, that the scientific category of 'Dinosauria' was first named. The man who described it, Richard Owens, a professor of the Royal College of Surgeons in London, was to be one of Darwin's fiercest critics.

It is the collection at the Museum für Naturkunde in Berlin, known for its wonderful original fossilized skeletons, that I think about most, though. This neglected museum, with its horrible plumbing and dank rooms, was starved of money for decades through the world wars of the last century and while it was part of East Berlin. It draws visitors now for its famous specimen of archaeopteryx, the ancient missing link between dinosaurs and birds, one of the fossils dear to Currie's heart.

But it was the *Ichthyosaurus communis*, unearthed in Lyme Regis, England, in 1846 – thirteen years before Darwin published *On the Origin of Species* – that caught my attention. The ichthyosaurus was a marine reptile that didn't survive the Cretaceous extinction. It had a long toothy jaw, four flippers, a tail fin and a body built to cavort in the sea. Think Jurassic dolphin.

The Museum für Naturkunde has a partial specimen, labelled with a polished bronze plate neatly affixed to its wooden cabinet. This is not a dusty bit like the pieces Currie has found for me. It is glossy, deep caramel brown. It seemed to me that I could see every fine rib, every bone in its back, every tooth in the long snout.

The secret inner workings of the flipper were on display too, laid bare by the process of fossilization and liberated from its resting place in the earth. I felt as if I could see not only inside the body of this marine reptile that had lived so many millions of years ago, but also deep into the past, long before the merry forces of nature had come up with humans.

It's an unaccountable skeleton if you're a Victorian scientist trained in the fixity theory. What must those earth-diggers who discovered this specimen in the Lyme Regis of 1846 have thought? How could they have reconciled it with the legends of their day? They didn't just write it off and discard it, because here it is, more than 150 years later, displayed in a museum. There must have been space in the legend, some flexibility that let them understand that it was beyond their orthodoxy. Maybe it was the same space that allowed Darwin to form his theories in the first place, however reluctantly.

I seek out Currie again, now confined in his office, surrounded by excavated fossils. I have looked at his theory of the fifth extinction, across the broad sweep of glaciers and continents, through deep time, into the very bone beds and fossilized spores of the K/T boundary. What story are the fossils telling us now that we aren't listening to, poised as we are at the edge of a sixth torrent of extinction of our own manufacturing?

It's that we are not exempt from the forces of evolution, he says. The legend of our society tells us that we are the pinnacle of evolution. And we act as though evolution no longer applies to us. Our population has risen to 6 billion and counting. We have killed off many of our natural predators, defeated many diseases, dramatically lengthened the life span of individuals in the richest countries. The irony is that in the process, we may have shortened the life span of our species on earth. It's the polar opposite of the way natural evolution works. Evolution preserves the species at the expense of the individual. This human-forced evolution prolongs the life of individuals at the expense of the species *Homo sapiens*.

The clearest sign of that is the loss of so many species of plants and animals that are part of the planet's genetic storehouse. It's most obvious in Madagascar, but is also happening on every other continent and large island around the globe.

A palaeontologist can see that the life forms that are disappearing – either going biologically extinct or vanishing in numbers and in areas large enough to matter – all work together to provide a system of life support, a broad genetic safety net against disaster. Now, humans are tampering with the chemistry of the atmosphere and therefore global climate. In the multi-billion-year history of the planet, this qualifies as catastrophe. It is classic ecosystem collapse. It makes our species and many others less resilient. Vulnerable. Like the dinosaurs.

Look, Currie says. Lots of mammals were around during the reign of the dinosaurs. At the end of the Cretaceous, the first primates had already evolved. They were nervous night-feeders that lived in trees and were much like the modern shrew. Pretty much inconsequential in the grand scheme of things. But they made it through the K/T boundary. The same evolutionary forces that propelled them from inhabiting the ragged edges of an ecosystem to dominating the world's ecosystems are still in play. They have to be. We're just too unpractised in thinking about deep time to see it. We tend to think what happened yesterday is the most important thing on earth.

The ancient bone beds have a story to tell us not just about the past, but also about the future. The trick for us is to be willing to hear the tale, to be willing to understand that evolution did not stop with us. Can we read into them not only what is convenient, what reinforces what we think we already know, but also face up to uncomfortable new ideas? Not only the modern equivalent of proof of Zeus's griffins, but also the culturally unknowable. That's a chance the dinosaurs certainly never had.

Currie has the odd palaeontological sense of humour that comes from understanding that almost every life form that ever existed has gone extinct. It's a gentle fatalism born of the evidence that extinction is not just the agent of destruction, but of renewal as well, both necessary as well as capricious.

He leaves me with a final thought. Like the dinosaurs, we are in the process of creating a fossil record that might be read in millions of years. What will it show? he wonders. An extinction spasm as big as the one at the K/T boundary? There's a good chance that's exactly what's going on. And if we know that's even a possibility, he says, shaking his head, wouldn't it make sense to do something about it? If, that is, we still have time.

FOUR

Parched Oasis

Jordan is a land of parables both ancient and modern.

This arid country, the meeting place of Africa, Asia and Europe, is the birthplace of modern civilization, home to the sacred sites of the early believers in Yahweh and Allah and the site of Jesus's baptism. It was a key holding of the Roman Empire, heartland of the medieval crusades and centre for the intrigue and betrayals of Lawrence of Arabia during the First World War. Now known as the Hashemite Kingdom of Jordan, it is a small, highly cultured power broker in the agitated Middle East whose King Abdullah is a direct descendant (by thirty-nine generations) of the Prophet Muhammad.

Jordan's modern parables are just as fascinating as the parables of old. I have come to bear witness to three of them. Now, though, they are ecological, and they are told with precise lessons for other parts of the planet. Each is inexorably tied to the desert and to the fact that geography, history and politics have conspired to make water the main currency of this region.

The main character in all of this is the heat. It is the dominant physical presence; it surrounds, embraces, refuses to let go. It has a texture – a personality. Life gathers only where there is shade. Trees straggle. Grass is absent. A flower is a gift, not of the omnipresent heat, but of the water.

The first two stories I am pursuing are billed as international conservation victories. In fact, I am travelling with a group of biologists from the World Conservation Congress, which has been meeting in Amman, on a tour designed to do a bit of gentle bragging. At 7.55 on this fall morning it is already 25°C.

We are heading deep into the desert, a black desert born of basalt. It is as flat and featureless as the surface of the ocean. No drop of rain has fallen here for four years, not even the 14 millimetres or less that the people can usually count on in a year. Occasionally we spot a desert castle in the distance, some built by the Romans nearly eight hundred years ago, used by the Byzantines and later transformed into sumptuous hunting lodges by the caliphs of the eighth century.

We arrive at the Shaumari Reserve. The stark sands are not so much grain as powder. Each footstep sends a puffy cloud above the ankles, where it hangs in the blistering air before settling down again. The fine grit seems to combine chemically with body salt to form a new, intractable substance. It is impossible to remove from clothes and shoes even months later. A wheatear flies by, flashing its white rump and black tail, the first bird we've seen in the couple of hours since we set out.

This is the centre of the universe for the Arabian oryx, a large desert antelope with long straight horns and a glossy white coat. It is said to be the prototype for the magical unicorn, bestower of never-ending life. To Arabs, it is a talisman of dignity and courage, a fierce fighter that evolved in the harshness of the Arabian desert. For millennia, the oryx has been a legendary hunting trophy, caught only after a bloody hand-battle between man and beast. It is the worthiest foe in the desert.

Lee Talbot recounts all this, standing next to me at Shaumari. A slight, white-haired man, he was the first staff ecologist of the World Conservation Union (IUCN), the biggest and most influential conservation organization in the world, and its director-general in the first part of the 1980s.

In 1955, he tells me, when ecology was in its infancy and the science of conservation was embryonic, the IUCN sent him on a mission to the Middle East to survey the state of the Arabian oryx. He could find fewer than sixty survivors, a number that meant the desert antelope was perilously close to extinction.

The reason for the collapse of the population was clear. Instead of tackling the oryx in the time-honoured way on horseback with a spear, the Arabian hunters had begun stalking the wild beasts in Jeeps or hunting them from aeroplanes. The oryxes didn't stand a chance. They were gone from Jordan in the 1920s. The last wild oryx was killed in 1972 by a hunter in Oman who wanted it as the ultimate trophy.

But there were still a few left in zoos around the world, and Talbot had earlier managed to capture three in the wild to use as breeding stock. He organized a worldwide rescue, taking a total of eight of the zoo animals and the three breeders to a spread in the deserts of Arizona. Once there, absolved from the need to dodge bullets, the survival herd reproduced mightily.

In the late 1970s, 11 Arabian oryx were returned to this reserve in Jordan. By 1983, there were 31. Today, Shaumari has 142. This Arabian nucleus of oryx production has exported 26 to the United Arab Emirates and 12 to Saudi Arabia. That's a total of 180 oryxes reintroduced to their native ecosystem. It's a resounding success in conservation terms. An eminent quarterly journal of conservation has been named *Oryx* in honour of the fact that this is the world's first victory in bringing back a big animal from the edge of extinction. Talbot is proud of that, even now, fifty years after he first sounded the alarm.

When we hop into trucks and drive around the reserve's dusty 22 square kilometres to find the oryxes, it's easy to see why. Most of the whole world's population is right here in front of us – 142 of them – their white coats shimmering. They look like mythic creatures against the heat-shuddering horizon. We keep our distance. The windowless truck containing a dozen humans doesn't faze them at all – if we get too close, they will attack with all their storied ferocity. Their horns can slice through a truck's metal as fluidly as my hand waves through the parched air.

They are grazing, heads down. But they are all too aware of us. Their ears flick. Then their tails. We edge closer, and I realize that their bellies and faces are slashed with black and they have black stockings. They are oddly proportioned. The horns that give them their name (from the Greek *orux*, a stonemason's pickaxe) are longer than their legs, honed to a point and lethal.

The oryx is a marvel of evolution, perfectly suited to this extreme climate. It needs so little water that it can live off the scanty dew of the desert. Talbot has a record of one oryx that lived for twenty-two months without a drink. They have also developed massive feet, the perfect evolutionary tool for walking in the dusty sand without sinking, like the caribou of the far north, which need spreading feet to walk over the snow.

By contrast, humans seem woefully ill-adapted to these surroundings. Many of the people in my group are looking distinctly wilted after just a few hours in the desert heat. All of us are caked with sand and sweat. Our guides take us back to some buildings at the entrance to the reserve and give us cup after cup of hot sweet tea. It's an ancient desert trick to make the body think it's cooler than it really is. The last thing you should do in heat this intense is drink something cold.

For the first time since we arrived, flies start buzzing around. I can't figure out why, and ask one of my scientist companions who lives in the Middle East. That's easy, he says, shrugging; we're in the shade. I still don't get it. He points to a bead of sweat on my arm. In the high sun, that would evaporate right away. Here in the shade, it stays around for a moment or two. The flies are landing on my skin to drink my sweat.

All this the oryx could survive. In fact, it thrives here. It was only human pride of conquest – and modern tools of destruction – that it could not withstand. True, human ingenuity has snatched the oryx back from genetic obliteration for now. But that was only by chance, a brave experiment that happened to work. A mere 180 of the species may be trumpeted as a conservation victory, but it is nowhere near recovery or even a permanent solution. The gene pool is shallow and therefore inelastic. That makes them vulnerable to disease and the ravages of an erratic climate. More critically, the oryx's place in the ecosystem, its biological efficacy, has vanished – along with the ecosystem itself.

Once this was a rich and vibrant desert system, with blue-necked ostriches, gazelles, Asian wild asses known as onagers, jackals, cheetahs, wolves, red foxes, the highly poisonous black Egyptian snakes and the now extinct red-necked Syrian ostrich. That assembly of animals – not to mention the plants, insects and other life forms they supported and that supported them – is gone now, dismantled piece by piece. The 6,000-square-kilometre national desert park, Jordan's first national park, has been dismantled too, under the pressures of human demand.

Today, all that's left of that large protected area is this 22-square-kilometre oryx reserve and the 12 square kilometres that contain the Azraq Oasis, one of the world's most famous wetlands. Species are going extinct here at the rate of about one a year, and have been doing that for the past century. That's way too much, too fast. In normal times, when extinction is a trickle and not a rush, it might be one or two species in the world over many centuries.

It is sobering, this tale of the oryx. This near-extinction occurred not because there was a struggle for survival between humans and animals but thoughtlessly, for pride of conquest. The oryx heads the hunters bagged were little more than baubles to hang on the wall, collected without an understanding of the great survivor's place in the ecosystem or that its absence would have an effect on other species.

You could say that it happened long ago, before anyone really understood that there were ecosystems and that they could be damaged, and that we know better now. The hitch is that we continue to do the same thing, even as our understanding of ecosystems and extinction becomes more sophisticated.

I can't help thinking about apes and monkeys, our closest genetic relatives. They are so seriously endangered now all over the world that conservation groups have recently made an anguished plea for $25 million (US) to stave off the extinction of several species. The groups say that the money – if they get it – will not prove a cure, just the equivalent of offering bread and water to a dying man. Just enough to keep some species from vanishing from the gene pool for now.

And what's killing the apes? Humans. Either outright, for the meat, or indirectly by taking over the places the apes need to live. It's similar to the state the oryx were in fifty years ago, except that apes are much harder to bring back from the edge.

What is our future if we keep doing this? What species will remain on earth if these tenacious creatures cannot survive? Apart from the ethical questions about whether humans have the right to use the planet this way and the aesthetic ones about what the world will be like once all these magnificent creatures have died off, there are the purely selfish ones. What if human survival at some point in the future depends on these species we are so cavalierly driving into extinction?

All these questions run through my mind as we board our bus again for the short trip to the Azraq Oasis a few kilometres away. Again, this is being pitched as a Jordanian success story: the fable of an age-old oasis destroyed and then miraculously resurrected. It has been written up in international documents as a show-piece of water restoration.

Water is a magical thing in the desert. It is the sought-after, the beloved, the unattainable. The water at Azraq has been even more. It has been the cradle of life in this vast Syrian desert for 250,000 years, the only constant source of water within 12,000 square kilometres.

Palaeontologists have found the lower jaw of an elephant from a quarter of a million years ago in the Azraq wetland. It's thought to be the missing link between the modern Asian and African elephants. They know that gazelles used to drink here thousands of years ago, that lions and cheetahs stalked their prey on the edges of the oasis among the riot of water plants. Hundreds of thousands of birds used this as a resting place on their migrations up and down the continents.

Humans have been drawn to this piece of lushness in the desert since the early Stone Age. Archaeological evidence suggests that this watering place was key not only to the desert's nomads, but also to some of humanity's first agricultural settlements. Azraq is so important to the hydrological machinery of the planet that it has been designated a Ramsar site under the global environmental treaty on wetlands.

It's a lucky geological accident. The area under the oasis is made of highly porous – and ancient – limestones and sandstones deep within the body of the earth, surrounded on top by a basin. That basin, like the cupped hands of a giant, collects the meagre rainfall from thousands of square kilometres and funnels it into the porous rock. Tiny pockets within the rock collect water, purify it and store it as massive aquifers. The whole process can take thousands of years. In turn, those underground aquifers feed springs that flood the surface.

Azraq was blessed with ten of these springs, which created a marshland of over 800 hectares, an embarrassment of aquatic riches in the desert. Of these ten, the most important was the mighty Ain Soda Pool, the spring that lay at the ecological heart of the oasis. For a quarter of a million years, the Ain Soda poured millions of litres of water across the marshlands every day. It was the ultimate renewable resource.

I've read the Ramsar report about what went wrong here and how scientists have fixed it. I'm weary of half-victories and outright failures and am anxious to see something that humans have done right. I want to believe that human ingenuity can overcome even the most horrible of obstacles.

The instant we get off the bus at Azraq, the air is transformed. It smells fervid and promising. Expectant. For the first time since I arrived in the Middle East, I can smell water.

There's a visitor's centre and a raised wooden path winding a short way through the reserve over the water. For a while I'm lost in the delight of finding anything wet. Red dragonflies play over the surface of the marsh, darting and then hovering. I can hear a mud-skipper splashing delicately over the surface. A guide explains the miracle of how the fish came back. It turns out that some species of fish eggs can survive being buried in dust for years. When the water comes back, they mysteriously rehydrate and hatch. It's the same story with rhizomes. Wizened by heat and drought, the water made them plump again and brought them back to life.

Handfuls of birds have returned, drawn once again by glinting water to this essential link in the international migration route. It's again possible to see purple herons and Egyptian nightjars and laughing doves. The honey buzzard flies by every now and then. The steppe eagle is back.

But the water is shallow. There is just enough for minnow-sized fish to swim. I can see ripples on the surface as they pass by. The guide tells me that when the oasis was healthy, the water would have been up to my neck. I keep expecting to turn a corner and see more water, more oasis. I've read that this wetland was 800 hectares, and I expect to spend hours wandering around here.

The wooden path leads me to a widely cracked depression. The educational sign posted above it is blunt. This is all that is left of the Ain Soda Pool, the enormous spring that kept the oasis ecologically alive for a quarter of a million years. It cannot come back to life. This is a gravestone in all but name.

Ain Soda died in 1993, just thirteen years after the water authorities in fast-growing Amman began pumping water from the underground aquifers that fed the spring. In the first year, 1980, they took out 15 million cubic metres of water from the Azraq basin. That's just short of the 20 million cubic metres or so that the Jordanian government's scientists believe can be replenished naturally in a good year for rain.

By 1981 authorities had dug fifteen artesian wells to the northwest of the oasis, the better to extract purified groundwater for the thirsty citizens of Jordan.

Ten years later, in 1991, the water-mining operation had risen to 39 million cubic metres a year, nearly double what the basin can sustain. As well, the Jordanian government had developed its main military airbase directly to the southwest of the oasis. It's unclear how much water the base extracts, but I can see lots of action there.

By 1993 the oasis that had sustained life for a quarter of a million years was a dustbowl. The springs dried up. Deep cracks broke open in the earth and spread into chasms. Plants died. Migrating birds abandoned it. Palaeontologists, making the best of things, started some excavations.

Eventually, fire replaced water. The soil underneath the oasis, rich and peaty after hundreds of thousands of years of being covered with water and vegetation, caught fire deep underground. I can still see the residue of the long-burning fires in the blackened soil. This is Darwin's theory of evolution in action. Change the conditions and species will change. In this case, the conditions changed so fast that species died off. Something may come back, but if it does it will come in at whatever leisurely pace evolution chooses, over thousands or maybe millions of years.

Water is laden with meaning in human culture. In a purely scientific definition, it is a combination of hydrogen and oxygen, a liquid essential to life, medium for nearly all chemical reactions in living things. In the metaphysical sense, it is the prerequisite for life on earth and possibly the origin of life itself. In the ecclesiastic sense, it is a powerful symbol of cleansing, renewal, acceptance and rebirth. It is nothing less than an emblem of faith in the future and of hope.

But just as water has meaning, so does the absence of water. It is the lack of life, of the possibility for life, of the potential for reaction or change or survival.

In the most basic possible sense in the case of Azraq, it also means that not only are the non-human life forms being sacrificed for the water taps of Amman, but the future of the pumped water is at great risk too. As the fresh water is relentlessly mined from the ancient aquifers, it leaves a space that salted water must fill, according to the dictates of nature. The 1990 Ramsar report said it is only a matter of time before the water from Azraq that sustains human life in the desert will be unfit for humans to drink.

Azraq is far too powerful an icon – and too critical a piece of the waterworks of the planet – for it to perish so ignominiously, so quickly. The government of Jordan and the international conservation community rounded up some money to mount a recovery. By 1997, this small portion had come laboriously, expensively back. It's roughly 10 per cent of the original oasis and is still seriously out of whack ecologically. Reeds are the most common plant, a weed species that shoves out the rich array of plants that once grew here. And yes, there are migrating birds, travelling the routes from Asia, Africa and Europe, but nothing like the 347,000 spotted by counters from Jordan's Royal Society for the Conservation of Nature during an aerial survey in 1967, before the serious pumping began.

The 1.5 million cubic metres of water used for the recovery is from fossil water even further down inside the earth than the original springs that once kept Azraq fertile. It's not clear how long the Jordanian government – fiercely proud of Azraq's place in world ecology but under great pressure to keep up water supplies in Amman – will be able to justify pumping the fossil water back into the oasis.

Worse still, the pressures on Azraq's underground aquifers are as strong as ever. At the beginning of this century, officials were admitting to pumping 37 million cubic metres out, almost double what the aquifers can replace naturally. By 2002 Hazem al Nasser, Jordan's Minister of Water and Irrigation, said that 47 million cubic metres were coming out of Azraq. He made an announcement early that year that the government was aiming to reduce that to 45 million cubic metres.

This is more than twice what the oasis can stand. One glass of water out of four drunk in Amman still comes from Azraq. As well, farmers are pumping aquifer water from five hundred illegal wells to keep their crops alive.

And there's a bigger problem. Even if the Jordanians can justify the financial and ecological cost to keep the sliver of oasis wet, the 1.5 million cubic metres directed to this restoration is fossil water. And, by definition, fossil water is finite. At some point, it has to run out.

This is not the promised resurrection. It's a sleight-of-hand, a clumsy, temporary man-made wetland that doesn't begin to mirror the elegance of the oasis that time and nature made. And it's alive at all only on sufferance. Any one of a list of changes – war, even greater population growth, a change of the Jordanian government or king, a change of heart – could see this restoration dry out too.

I keep wondering if I'm seeing it properly. Can it really be this terrible? Wasn't this supposed to be an international conservation victory? I consult with yet another eminent scientist who has been with me on this excursion. He is Roger Crofts, a geomorphologist who is the chief executive of Scottish Natural Heritage in Edinburgh. He's seen exactly what I've seen over the past hour and more. Right now he's in the visitors' centre, reading a chart on how much water is still coming out of the aquifers. I sidle up to him and quietly ask: Can this last? No, he says. The Jordanians are using up their environmental capital. Once it's gone, it won't come back. Or, at least, not in a comprehensible human time frame.

It's not just Jordanians who behave as though the earth's water is inexhaustible, who think, in the Malagasy way, that if they keep pumping there will always be another litre to drink. The same story is being told on every continent all over the planet, even in places where water is not considered to be scarce and where water conservation is unimagined. Among the most endangered aquifers are some in the United States, Mexico, India, China, Pakistan, parts of Europe, Africa and other parts of the Middle East.

In fact, compared with the citizens of other countries, the Jordanians scrimp on water, using only about 85 litres a day on average. In water-parched Syria and Lebanon, it's 125. The richer Saudi Arabia and Gulf states have a relative gush of water with about 400 litres for every citizen every day. In North America, where water is so plentiful few citizens even think about where it comes from, the personal use every day is between 500 and 600 litres.

Despite the seeming plenty, though, water is turning out to be a problem even in North America. Deep beneath the eastern flank of the Rocky Mountains, stretching through South Dakota down to the Texas Panhandle, lies the Ogallala-High Plains aquifer, one of North America's great bodies of underground water. The aquifer has fed agriculture in this area for about 100 years, defying the predictions of early explorers that this desolate plain could not support human life. Human ingenuity came to the rescue with irrigation, pumped out of the aquifer. Now, farmers in eight states covering an area of 480,000 square kilometres grow wheat and other grains in the world's largest irrigated expanse of cropland.

It's taken longer than the thirteen years at Azraq but, a hundred years on, the High Plains aquifer is feeling the effects. By 1980 the water level beneath the Texas Panhandle had fallen more than 30 metres since irrigation started. By 1998, just eighteen years later, as irrigation stepped up, it had dropped another 12 metres. A shocking 1.5 metres of that drop came during 1997 and 1998, an unusually wet period when the aquifer ought to have had lots of scope to bring its levels up.

It seems to me, standing in the Jordanian desert, that humans are so busy making ingenious technical advances for our own generation that we are forgetting the big picture. If we use up this much water now, in times of plenty and relative peace, how will the humans of the future make do? Aren't we in danger of shortening the shelf life of our species – and that of many, many other species – in the name of living out this fable of inexhaustibility? Is this not a sister belief to the idea in Darwin's day that species were fixed? Except now we understand that species can go extinct and have gone extinct. But we don't seem to extend that to understanding that ecosystems, that life-support systems like aquifers, are not fixed. The desert of Jordan, the vanishing forests of Madagascar, even the catastrophic changes that seem to have affected the dinosaurs, show that ecosystems adapt too, and not only in the ways humans want them to. Ecosystems, like species, like our planet in the galaxy, are not fixed.

I can see human goodwill at play here in Jordan. The oryxes have been brought back from absolute extinction. Azraq is no longer a smouldering pit. But I have to ask myself – why is this the way we use our intellect, bandaging woeful problems instead of preventing them? It has to be that we don't understand the powers of evolution. We understand that evolution formed species in the past, but we don't seem to accept that it will also make them change in the future. Species that have adapted over time to make this incredibly rich and varied world are programmed to continue to adapt if they need to. If we force their environment to change, they will change too, even if it means dying out.

I head back to Amman in a funk, thinking of ancient Sumer, the world's first catalogued civilization. It thrived not so far from here six thousand years ago, when the technologically adept Sumerians figured out how to harness the power of the Tigris and Euphrates rivers to water their crops in the rich Mesopotamian soil. They prospered, growing wheat and barley. Sumer came crashing down after two thousand years when the soil, flooded and dried out time and time again, became poisoned with salt and other substances.

Yet the Sumerians, like the Jordanians, the Malagasy, the Americans and all the civilizations where the fable of inexhaustibility is playing out, were not thoughtless. Their society was not defined by destruction alone, but also by the fluted golden cups they are still famous for. For the enchanting bowls of black obsidian, the imposing ziggurats said to be the architectural model for the Bible's tower of Babel, for the political template of the city-state, the money economy and cuneiform writing, the foundation of many written languages.

Humans have evolved not just to destroy, but also to create, not just to see the here and now but also to imagine the future. I think about the New York Museum of Natural History and its collection of some of the earliest human-made sculptures yet found. They are lithe and elegant animals, small enough to grasp in a single hand, 32,000-year-old talismans from a species capable of understanding symbol.

This is a prodigal era. There are no pioneer lands left to break, no Mesopotamian deltas left to poison, no vast Malagasy forests left to cut, no new North American aquifers to drain. There is nowhere else to go once we bungle life on this earth.

THE TALES OF THE ARABIAN ORYX and the Azraq oasis are powerful international symbols to the scientific world but they pale next to the mainstream celebrity of the subject of the final Jordanian parable I am to examine: the Dead Sea.

I journey there by night. Secretly, I am hoping that the sweltering heat will lift by the time I get there from Amman. Wrong. It turns out that the Dead Sea, nestled between Israel and Jordan, is even hotter than the rest of this country, the result of the strange microclimate that helped create the sea in the first place. It is a sauna.

This is the lowest point on earth, a dead-end sea at the bottom of the Jordan River that fills up a rift made perhaps five thousand years ago when two tectonic plates shifted and this little piece of the planet's crust collapsed.

It is one of the world's great natural marvels, salty enough that it holds up the weight of humans. Salty enough that few forms of life, save for a couple of bacteria, can live here. The saltiest body of water on earth. It's a phenomenon caused, of course, by Jordan's otherworldly heat. Over the millennia, as the Jordan River poured into the sea from the north, the water evaporated quickly into the atmosphere, transformed from liquid to gas by the temperature. Salt, though, doesn't evaporate and it was left behind. Over time, that caused the sea to turn to brine. The flow and the evaporation eventually formed an odd saline equilibrium that has fascinated scholars for thousands of years.

The Latin chronicler Tacitus mentions it in his *Histories*, written shortly after the birth of Christ. The ancient writers Pliny and Josephus talk about it. To biblical writers, it was the great lake of the Holy Land, referred to throughout the Old Testament, including several times in the Book of Genesis, the prophecies of Ezekiel and Zechariah, and the Apocryphal books of Ezra and Nehemiah.

The most famous story concerning the sea, contained in Genesis, is the story of Lot, which goes like this: Lot lives in the city of Sodom on the shores of what scholars now believe to be the Dead Sea. But it is a city marked for disaster because the men who live there lust after other men, strictly forbidden by the sexual mores of the day. As punishment, God rains fire down on Sodom, razing the city and killing all the inhabitants. Lot and his family are chosen to escape His wrath as long as none of them looks back to witness the carnage. Lot's wife can't resist. She looks back and is turned into a pillar of salt. Lot leaves her there and flees to the mountains, where he has sexual relations with his daughters and produces grandsons who head the blighted nations of Moab and Ammon, fated to be continually at war with the Israelites.

Surely the Dead Sea was a cursed place. Noted for its uniqueness, feared for its unnaturalness, it was a trope for death.

It's just as famous today, although for different reasons. Posh resorts and health spas have been built along both the Israeli and Jordanian sides of the Dead Sea. Europeans flock here to take an annual cure, letting the salt water drain away their cares.

Except that now the Dead Sea is dying. In fact, it is in danger of disappearing altogether. During biblical times, and until fifty years ago, the Dead Sea had a surface area of about 1,000 square kilometres. Today it is roughly two-thirds of that and is shrinking day by day.

Historically, it rested at 395 metres below sea level. In less than fifty years, it has fallen to 410 metres. Today, it is dropping by about a metre a year. The Jordan River, once so robust, has been relentlessly dammed and diverted. It used to pour 1,000 million cubic metres of water per year into the Dead Sea. Today, it is a comparative trickle, only about 150 million.

Not only that, but both Jordanians and Israelis have begun harvesting potash and other minerals from the Dead Sea by using the heat of the place to create industrial evaporation pits. The water that's left over from this industry is brinier than ever and is pumped back into the Dead Sea, minus about half its volume. This, too, is shrinking the sea.

The result is dramatic. Health spas and resorts that were on the sea's edge a few years ago are now half a kilometre away. Israel's Ein Gedi resort, once right at the water, has taken to shuttling would-be bathers to the sea in a tram. Worse still, as the Dead Sea retreats, fresh water has begun to seep around its circumference, soaking up salt. This destroys the structure of the land, and it collapses into itself, creating massive sinkholes. One sinkhole on the Israeli side was so deep and wide that it swallowed several groves of date palms. Sinkholes are apocalyptic enough on the Jordanian side that three thousand residents have been forced to flee the disaster.

There's still some of the sea left where I am, further north on the Jordanian side. Here, rather than expanding shores, there are deepening drops down to the water. I scramble down, eager to feel the legendary brine on my fingers, aware of the flashes of gunfire on the Israeli side. A new chapter in the long-running Middle Eastern war has recently started. Ceaseless gunfire, land collapsing under the feet, hellish heat. Maybe the Dead Sea curse is still in play, I think.

But this tale of doom is not the end of my final Jordanian parable. The Israeli and Jordanian governments have big plans to preserve this iconic site. In 2002, at an international environmental meeting in Johannesburg, they declared that the Dead Sea is in grave danger. They asked the world financial community for some help – roughly $800 million (US) to start – to build what they call the 'conduit of peace'. It's a 320-kilometre pipeline to take water from the Red Sea to the Dead Sea. The aim is to replenish the waters of the Dead with the waters of the Red and resurrect the Dead Sea. As the water travels from the sea-level Red down to the Dead, it will create energy, to be used to take the salt out of the Red's water and turn it into fresh water for the Jordanians and Israelis to drink. The brine would then be pumped into the Dead. Building this desalination system would cost an estimated $3 billion.

This is a new incarnation of a decades-old plan to

pipe water from the heavily ecologically degraded Mediterranean Sea, which is closer. That idea, known as the Med–Dead pipeline, has lost favour.

The ecological effects of the Red–Dead plan are unknown. Among the questions is what will happen when the two seas' waters are mixed, because the chemistry of the waters is quite different. As well, it is not clear what the removal of the estimated 1,900 million cubic metres of water a year from the Red Sea will do to its delicate ecosystem.

The governments have promised to look into it.

FIVE

Sunken Graves of the High Arctic

For those who live high above the Arctic Circle, survival depends on being able to read the code of snow and ice. And so they watch.

Talking about what they have seen doesn't come nearly as easily to the Inuvialuit of Banks Island; they are close-mouthed, especially among outsiders. But what they have seen is worth knowing: this is one of the first communities on earth where the effects of global climate change have been catalogued.

I have chosen to come here in the dead of winter, five days after my fortieth birthday, in the hope that the Inuvialuit will tell me what they have seen. It's not at all certain that they will. To them, I am a southerner, ignorant of the complex mysteries of the north. And because of that, perhaps not worthy of being told.

By chance, I sit next to Sarah Kuptana on the last leg of the journey to Banks Island from the south. She is well into her eighties, the eldest of the elders of Sachs Harbour, from one of the community's two oldest families. I slide in next to her, a slight and stooped woman bundled into the corner of her seat next to the window of this Beech 99 twin-engine aeroplane.

All ten passengers stuffed into the aeroplane are swathed in heavy winter gear. I'm in a goose-down coat on loan from a colleague, heavy boots that reach to my knees, two pairs of mittens and a hat. It's not enough. This is a two-and-a-half-hour flight over the Arctic Ocean in a barely heated aeroplane. My knees are clamped together, my hands stuffed under my armpits, shoulders hunched. I am numb, wholly concentrated on trying to keep warm. A southerner.

Beside me, Sarah is reading the ice of the ocean through the eerie afternoon twilight, silently, intently, almost greedily. She has been away for a stay on the upper edge of the North American mainland and has missed it. For part of the journey we are only about 15 metres above the ice. As we near Sachs Harbour, she shakes her head, clacks her tongue and points. Below us, the sea ice looks like a sheet of frosted glass that has been dropped on a concrete floor. Parts of it are shattered. Between the shards, great dark stretches of ocean are visible. This is not how it should be. This is the middle of winter. The ice should be thick. It should be solid. It should be dependable.

Local residents have gathered at the tiny airport, eager to see who is coming, who is going. A large party is meeting Sarah to welcome her back. Even inside, they wear huge parkas, big boots, hats lined with fur, mittens of hide. It is -35°C. The snow squeaks under my boots as I get off the plane. The air bites my lungs. In the time it takes to walk to the greeting room from the plane, my cheeks have no feeling.

Banks Island is one of the most remote inhabited places in the north, the westernmost island of the Arctic archipelago. Polar night – when the sun does not break the horizon for months on end – has just started to ebb. Now a slim red sphere of sun peeks over the horizon for a few minutes a day, tingeing the clouds with unearthly tones of pink and orange, before setting and plunging the day into darkness again.

This is part of the world's largest polar desert. Beneath the snow is a barren plain covered in frost-shattered rock and permafrost. And on top of the barrens lives the world's largest population of musk ox, perhaps sixty thousand, a majority of what's left on the planet. The musk ox is one of the oldest mammals in North America, a fearsome, long-haired, cloven-footed survivor from the Pleistocene that was a contemporary of the mammoths and the sabre-toothed tigers.

The ancestors of the hundred people who now live year-round in the hamlet of Sachs Harbour on Banks

Island travelled as nomads in this region five thousand years ago, after the retreat of the Ice Age, following the ice and the musk ox, learning how to survive on the edges of the humanly possible in the bitter cold.

To the current generations in Sachs Harbour, this winter is disappointingly warm. It's not just the sea ice that has gone wrong, breaking up and moving when it should be staying put. Many other parts of nature have gone strange too. The Inuvialuit of Sachs Harbour have been keeping track of all the oddities as part of an international research project.

Rosemarie Kuptana, Sarah's daughter, who is one of the Arctic's most important political leaders, is the driver behind that. For several years, she has been on the board of the International Institute for Sustainable Development (IISD), a research organization. When she heard what her people were saying about the changing climate, and then when she saw it for herself in the legendary spring of 1998, when the winter retreated in a bang in just three days, she persuaded the IISD to catalogue what her people were seeing.

It's an exercise in mining the memories, the oral histories of this community, the knowledge that has been passed down through the generations to ensure the survival of the people.

What they have charted is a sharp shift in the ecosystem. Barn swallows for the first time in memory. Robins. Salmon. Herring. Mosquitoes. All creatures that the Arctic freeze has kept out of this area for thousands of years.

The baby seals are skinny because the ice melts so quickly the mothers don't have long enough to fatten them up before they move on. Some of the little birds that should be forced south by the bitter cold have started staying over the winter. The houses are shifting as permafrost slumps, melting in the unnatural warm.

It's just not as cold as it used to be. The school closes only at -40°C. In the 1960s, it used to be closed all the time in the winter because temperatures routinely went under -70° with the wind factor. Now, it rarely gets below -35°. The school closes maybe once a winter.

TIME THIS FAR NORTH is approximate. It does not depend on where the sun is, or what the horizon has to say. The horizon is important mostly for what it reveals about the sea, not the sun. And so, when I arrive at the home of John and Donna Keogak and their children late in the evening, they settle down comfortably to talk for hours. John, one of the island's master hunters, sits on the floor, his sturdy back against the wood panelling on the wall, one arm resting on his knee. The couch is for visitors.

The problem, John says, is too much water. It's happening all over the north. The north is not used to so much water. He talks about the early 1970s, when he first came to Banks Island and started hunting polar bear. The hamlet had only six houses and no electricity. Everyone drove dog teams, not the snowmobiles and trucks they have now. And the ice in the bay in front of his house went on for ever. No limit. And solid. You could stay out and camp on the ice while you hunted.

Now, the ice stretches only 6 or 7 kilometres. And what's there is unstable. You can't camp. Sometimes you can't even go out on the ice, it's so thin. He hasn't been out yet this year, even though it's already mid-winter. Now the ice is so shifty that he has to rely on more than his eyes and his experience. Now he has to take his GPS and a barometer and read them as well as the ice. When the barometric pressure drops, he gets out quick.

It's hard to predict the seasons too, says Donna. Last year, summer never came. Just a lot of wind, but no summer heat. And the wind at New Year's this year was so strong that it picked up the frame of their summer tent and moved it a good 10 metres or so. It had been in the same spot for nearly twenty years. Nothing inside moved. The whole thing just picked up and landed again like magic.

The seal hunting has gone strange too, says John. This summer for the first time he couldn't get enough to feed his dogs.

And the permafrost is giving up its secrets. He and some other men started finding a strange dark wood up to the north of the island. Finding wood at all so far above the tree line is rare – usually it floats here from the mainland if it comes at all – but this was odder still. They set it out, and when it was dry it burned like coal. Eventually they learned that it was petrified wood from eons ago when this was a tropical island instead of a polar desert. It had been buried deep in the permafrost for thousands upon thousands of years. Until now.

They know this is global warming. They know that the Arctic, their home, is fated to be affected first and worst by global climate change. That's how it works. This great global experiment will play out on them first. Donna feels sorry for the elders. They're just lost. They want to go out on the land, to dance the rituals of the seasons. They don't know what to do.

John shakes his head. He's sort of lost himself, he tells me. He wishes he'd been born thirty or forty years earlier, when things were predictable. When this corner of the planet was the same as it had been for generations. Now, there's no turning back, he says. He's been to Ottawa to tell the politicians about the changes in the Arctic. But that's not going to make a difference. The rest of the world isn't going to stop what they're doing just because his life is changing.

He wanders over to the kitchen, stretching his long legs at last. On the counter a long length of musk ox backstrap is thawing. Islanders have been allowed to hunt musk ox for food since the 1970s. It almost went extinct in the early twentieth century, an ignominious fate for an animal that had once roamed as far south as Kansas and Kentucky, once lived in England, France, Germany and Russia, and had survived the Ice Age without missing a step.

But it was hunted ferociously for its pelt and for meat once the bison had been shot out further south. It disappeared even from its stronghold on Banks Island in about 1914, not coming back in strong numbers until the 1950s, after the heavy hunting stopped.

Now most of the musk ox are confined to the far north of Canada and Greenland, with a few restored to Alaska. The islanders hunt it judiciously. They figured that out a long time ago. You don't take too much or there won't be any left.

John shot this one last weekend about 30 kilometres out of the hamlet, and he's getting ready to make this piece into a stew, or maybe fry it up. It's not the choice meat. Kids in Sachs Harbour would rather have ham or beef from down south, but that costs a lot of money. Getting a musk ox just takes the skills of a hunter. John doesn't say it, but I sense that he's proud he can feed his family by his own wits. At least for now. I can't help thinking about what will happen to those few remaining musk ox when it gets even warmer here. Much warmer.

THE INUVIALUIT are trained in subtleties. Where the uninitiated look at the Arctic landscape and see numbing sameness, the Inuvialuit read critical differences. They can tell when a storm is coming up, where the cracks are in the ice, how to get home if you're caught in a blizzard. They know the difference between the ice that stays around for years on end and the ice that just formed this winter. Old ice has a slight bluish tint, they say, as if something is reflecting up from underneath. New ice is paper white.

Roger Kuptana, Sarah's son and Rosemarie's brother, learned at twelve how to read the snow. His father, William Kuptana, taught him to cover up a trap with snowflakes and then sprinkle older sugar snow over the top to fool the animals. To make a house out of snow, you use young, fine snow packed hard by the wind. That way, the wind won't penetrate as much when you're sleeping. You use a snow-knife with a long blade to cut the blocks. When it's done, maybe forty-five minutes after you start, the snow house is so strong that you could stand on top. For tea, you use crystallized snow.

Roger has translated some of those skills into earning a living as a hunting guide for southerners. He and his wife, Jackie, an Englishwoman with a taste for adventure, and their teenagers run a lodge in Sachs Harbour to house the hunters and anyone else who visits. What most of the southern hunters want is to bag one of the 1,200 or so polar bears that live on the island and the ice surrounding it. People in the community of

Sachs Harbour are allowed to kill about a dozen of the sleek white bears in total every year for their own purposes, and to guide wealthy outsiders by dogsled to kill the same number. It's not an easy job, killing a polar bear, especially a big one. They're smart, says Roger. And the smarter they are, the bigger they get. But their meat is good, he says. He likes them roasted. They have a piggy taste to them.

Like John Keogak, Roger has been hunting musk ox this week. He shot ten yesterday in three hours. Even though they look massive, they have only about 70 kilograms of meat on them. He feeds them to his dogs.

I am staying in Roger's lodge, a modern building, huge by the standards of the far north. The logistics of transporting all these building materials – mostly wood and massive picture windows – from the south boggle my mind. It's about ten in the morning and the sliver of sun has not yet risen. I feel compelled to sit and watch the sky, to see if I too can learn the code of the north. It's like perpetual twilight in the east. The oddness of that doesn't hit home until I realize I've never seen twilight in the east, only the dawn. The horizon is razor-sharp in the east and the ice is so deep purple it's almost black. In the west, the ice and the sky merge to the eye of the newcomer and it seems there is no horizon at all.

The few streaks of light, combined with the electrical street lights, show me the fifty or so houses in the hamlet. They face the ocean, perched on the banks of the sea. Snow – I don't know what kind of snow – clings vertically to the sides of the houses, swept there by the wind, and falls off in brittle sheets. It's not just the light I miss, but colour, which is a gift of the light. Everything seems monochrome, pallid, asleep. I wish I could believe this was immutable, but what I've seen here makes that impossible.

THIS IS A SOPHISTICATED community with layers of secrets, decades of mystery. Sarah Kuptana is the keeper of the secrets, of the legends of this place. She has been watching me wandering around her land, asking questions, trying to understand. She has seen me learn to watch, to listen. She has decided to tell me part of what she knows, to explain why the people of Sachs Harbour, unlike so many others in the world, understand what's happening when they see it in front of their eyes. Why they are fatalistic about it.

I arrive unannounced, as is the custom here, breathless from the cold. Sled dogs that live outside no matter what the temperature have seen me walk across the hamlet and are setting up a deafening howl, one of the few things to be heard here above the wind.

Sarah is sitting cross-legged on her dark-green couch watching television in trousers and moccasins. Her sister Edith Haogak, who is in her seventies, sits next to her, hands clasped in front, resting on her belly. A tiny child, perhaps two or three years old, a young member of the Kuptana clan, is sitting on the floor. On the low table in front of them, next to the remote control, is a hunk of raw flesh that Sarah has been eating, perhaps seal. The house carries the dark scents of old blood and fish.

She speaks Inuvialuktun, the ancient language of the western Arctic, the language of her parents and grandparents. It is full of soft clicks, as if the tongue is drawn to caress the soft flesh of the mouth's palate. It is a richly descriptive, emotional language that only a few people on earth still know how to speak. She speaks English too, but what she has to say today must be said in the old tongue. These are the legends of her people. Peter Esau, a newcomer who has only been in this hamlet for forty-five years and who was once its mayor, is translating for me.

This is not cold, she says. Sixty below was cold. Then, they used to wear two layers of caribou fur. The first inside, the second outside. Her mother used to sew them for the children. All one piece. They didn't have a thermometer then, but they know now it was -60°C because the kerosene couldn't pour. They used to live on the ice all winter and travel like the animals, looking for food. They lived on seal meat. And when they caught the seals they would share them every time, even the blood. They would ladle it out to everybody for making broth. And they would never make fun of the animals. They would honour them. They would give the seal a drink of water after they killed it, to give thanks to it.

When Sarah and Edith were first married, they were the ones to go out first in the morning and read the sky, read the stars, tell the weather. The ice was their domain. They could look across the vastness of it in the half-light and tell where the cracks were. Sarah gave birth to Rosemarie, her last child, on the ice. Not on the land, but on the frozen sea. William, Sarah's husband, was out setting hooks for the seals. He came back in time to see Rosemarie born.

Sarah and Edith knew the sea would warm up because their father told them long ago that it would. He told everybody. Their father was a special man, a religious man, named John Kaolok. He had a gift from God, the sight, they called it. He said the people of the north would stop suffering, that it would be the people of the south who would go hungry. The cadences of Sarah's voice rise and fall like an incantation. She heard this long ago, she tells me.

The weather will change, her father said. The animals are going to suffer first. They are going to be gone. She heard this long ago.

Everybody knew. Everybody believed her father. Sarah's voice becomes deeper, clearer, more urgent. They believed her father because, once, a boy died, a little boy no older than that one, she says, pointing to the child watching television. And her father came upon him and began to pray. And the boy got up. Sarah takes off her glasses for emphasis and waves her hand in front of her face. She saw him come back to life with her own eyes, she says.

Edith nods in agreement. She too has heard this story. Many times. That little boy was Phillip Haogak. When he grew up, Edith married him.

Sarah has her own secrets, not just those of her famous father. Shortly after her father gave her to William Kuptana to marry, she got tuberculosis in her bones. Tuberculosis, an import from further south, was sweeping the north. William's first wife had died of it.

Sarah's bones were asleep, she tells me. They wouldn't move, not even her neck, just her hands. She was in the hospital for five years. The doctors took out eight ribs on one side. Finally, she was told there was no hope. She was all alone in the hospital. She felt as though she had already left her body and could see herself lying far down below. She heard an Anglican minister, and she began to pray. She lived. She still gives thanks to God for bringing her back.

Now her son Morris, born after she got better and came home to William, is an Anglican minister.

Peter Esau turns to me, finished translating for now. You see why they are really precious to us, he says. They know all these things. They are our last ones.

As we leave, I realize that Sarah has not touched the piece of raw flesh on her table the whole time we have been there. She has been so immersed in her tale that she has not even glanced at it.

*

MOST PEOPLE in the world never think about the Arctic, let alone consider why it's crucial to preserve this part of the world. Why does the Arctic matter? Some try to argue it on social grounds: the Inuit and Inuvialuit of North America's north are indigenous peoples whose way of life has intrinsic meaning. Others are nostalgic. They don't want this rich piece of the earth's human culture to vanish.

Yet others see it through the lens of science: the Arctic ice is the planet's regulator of temperature and climate. Tiny changes to the ice mean major changes to global climate. The snow and ice are like a hat on the top of the planet. They keep heat in the sea and land so it doesn't rise into the atmosphere. But when snow and ice melt, they can't keep as much heat in any more. It rises, heating up the atmosphere, which in turn melts more ice and snow. The melting feeds yet more melting.

Now the world's climatologists have said that increasing concentrations of greenhouse gases – such as carbon dioxide and methane – are contributing to the warming of the earth's surface and therefore to the melting of the Arctic ice. The gases act like an insulating blanket, trapping heat close to the planet instead of allowing it to disperse beyond the atmosphere.

Already, the northern sea ice has thinned dramatically. Scientists can tell by measuring up from below with nuclear submarines. They say it is 40 per cent thinner than it was forty years ago. And they can tell through satellite photography that it covers 14 per cent less area than it did twenty-five years ago. But that's not the end. The newest scientific information has found that the permanent Arctic ice is shrinking by 9 per cent a decade. Even if it held to rate and didn't melt exponentially, which is unlikely, the Arctic's permanent ice would be gone by the end of the twenty-first century.

This is not a trivial change. Some climate models indicate that if all the Arctic sea ice were to vanish, the heat of the ocean would warm the air above it by 20 to 40 degrees in winter. With the heat would come evaporation and then rain or snow.

That means that Sarah Kuptana's Arctic is likely to metamorphose into the place her father predicted: a warm and lush part of the planet.

This is happening so fast that the world's top polar bear scientists have suddenly started to wonder whether they are studying an animal fated to go extinct. They used to worry that too much hunting would hurt the population numbers. Now, they realize that the very habitat the bears need to survive is disappearing.

The immediate worry with the musk ox is parasites. In the 1950s the parasites would have a scant month to infect the musk ox before the cold killed them off. Now, it's three months. The problem is that the parasites dry up the musk ox's appetite. They live in the lining of the stomach and ulcerate it. Then they seep into the bloodstream and weaken the animal. What about when it gets even warmer? What will the parasites do then?

The Inuvialuit have learned to tread lightly on this fragile land, to use the resources sagely, to make them last. But no matter how sensitive they are to their place in this isolated fragment of the planet, it will never be enough to hold back the damage from far away.

As Sarah's father predicted, changes would not be confined to the Arctic. Looking at what's happening in this community near the North Pole – like searching for the vanishing life forms of Madagascar or witnessing the dying ecosystems of Jordan – is like peeking into the planet's future. Except that the ecological change happening here at the top of the world is coming not so these people can satisfy their day-to-day needs, but because citizens in other, richer parts of the world are pumping waste into the planet's atmosphere.

Already, scientists are cataloguing more erratic weather all over the globe, including flooding, droughts, heat waves and catastrophic storms. They are worried that humanity's ability to grow food and find fresh water will diminish.

Ranges for many plants and animals have shifted towards the poles. Plants are flowering earlier. Animals are breeding earlier. And the ice cover on rivers and lakes in the northern hemisphere has shortened by about two weeks. This is widespread, rapid ecosystem change, and it's not clear how life forms and life-support systems such as water will react.

Even if humanity stopped pumping these greenhouse gases into the air today, the earth's surface temperature and sea levels would continue to rise for centuries.

I talk to Rosemarie on the phone. She is on the mainland, just back from a meeting of the world's political leaders in The Hague on climate change. She showed them the video her people made of the changes they are seeing here on Banks Island. Now it's been shown around the world, but she's not sure it will help.

She would like to be optimistic because she too has children. But she has seen so much human bungling in her years. When she was seven or eight, she was sent from Banks Island to a Christian boarding school on the mainland. That was in the 1960s, when aboriginal children were taken from their parents to live at schools far away. Hers was like all the others, Rosemarie says. The children weren't allowed to speak their own language or wear their traditional clothes or play with their siblings. In fact, like the other children, Rosemarie was given a number to identify her. It was 475. She had to put it on her bed, her clothes, her boots. In class, though, she was called by her name. I can't help thinking of Darwin's Fuegian, Jemmy Button of the Yamana, and the attempts to 'civilize' him.

To her, global warming is in the same vein as residential schools: it's an intrusion into her people's land and culture. There has been the church, the state, the governments, technology, industry, business (and tuberculosis, I think to myself). Now this concentration of greenhouse gases could be the ultimate intrusion. The pollution of global warming, sent north by those who probably don't even know her people exist, is wrecking the climate, the very essence of how people in the Arctic define themselves.

PETER ESAU has decided I need to know what musk ox tastes like, and so he has invited me for dinner. He is a man of irrepressible energy and boundless good humour. He cooks the best musk ox in the whole area, he assures me. Secret recipe. When I arrive, he has been cooking all day, and the house is humid with the rich smells of cooking flesh. He pulls me over to the stove to show me the prize. His special ingredient? 7-Up, he chuckles, rubbing his hands together. He covers the meat in 7-Up and cooks it all day in a pot on the stove.

His house is filled with tropical plants, thronged together in front of his big living-room window, reaching towards the meagre Arctic light on what looks very much like an Ikea wooden shelf. There is a spider plant, a thriving fig, an umbrella plant and an ivy that he hasn't quite given up on.

He is a man who has striven for a dream, reached it and is satisfied. He has been here on Banks Island only since 1955, a relative newcomer. But he always wanted to come to Banks Island because that was where men made their fortunes hunting fox. It was the white fox capital of the world in those days. His father died when he was five, and he lived on the land with his mother and eight sisters. His mother told him to forget dreaming about Banks Island, to do something practical instead. But he came anyway, he tells me.

When he arrived, he didn't even know how to set a trap. He was a bit of a laughing stock to the other hunters, who had been taught at their fathers' knees. But the first hunting trip he made, he took more than a hundred foxes and the other hunters figured they were in for some trouble from the newcomer.

He worked hard. For five years, he saved every dollar he made. He had his own dog team. Eventually, he had the longest trap line on Banks Island. Later still, he made the world's record of the most foxes in one hunting trip: 317, he tells me. He's still proud.

And he has helped build this community. He still can't bear to be away from it for more than two weeks at a time. He was the mayor for a decade, and president of the co-op that brings in food and goods from the south. He and his wife, Shirley, have raised ten children here, four of them adopted. He has a big shed filled with meat he has hunted, fish he has caught. People know that if they're hungry, he will help out.

But the land has changed so much. It never used to rain, and now it does. He remembers hunting seals on the sea ice with his dog team on 17 July 1957. And on 1 July 1967, when Canada turned one hundred years old, they had dog-team races on the ice to give their country her due. That's how solid the ice used to be. The big jamboree to celebrate spring used to take place in the snow in the first week of May. Now it's slushy. And spring used to take a long time to come, used to tease the people of Banks Island, pretend to arrive and then turn her back for a bit and come later on. In 1998 spring came in three days flat. Everything melted, rivers and all. They couldn't hunt in the mud.

He's worried about the polar bears. He knows the bears. He's hunted them for nearly fifty years and watched them for just as long. The two-year-old bears aren't very good hunters. They live pretty much on seal pups. But he's been watching the seal pups too. They are born in the long ice ridges. And when it melts so fast, the ridges collapse and the seal pups die. What will the two-year-old bears eat?

Last year he tracked a female polar bear for weeks just to see how she was getting along. She was looking for some good deep snow to hibernate in. For weeks she looked and couldn't find any. If there's no snow, what will happen to the bears?

We sit down to his feast of musk ox. The musk part

is true, 7-Up or not. I keep thinking that this is the oldest mammal in North America, driven to the edge of extinction even here on Banks Island a few decades ago, survivor of everything but the guns of man. And then I eat. This is not the feast of dreams, but it is lovingly made and offered. I eat it in the same spirit.

ANOTHER FEAST is in the planning. The whole community is to gather in the school gymnasium to watch the video that IISD has made and that Rosemarie has shown around the world. Most people in the community haven't seen it. The feast will also honour Graham Ashford, one of the people who works with the IISD in the north and in India and who has spent a great deal of time in the hamlet over the past few years.

A feast is a big deal, especially at this time of year. Some of the women in the community have laboured over it for days. And there has been much hand-wringing about how to pay for the food, which is horribly expensive so far north. Graham has a few hundred dollars and I help out. It will be just enough.

First, though, I want to see where the people of Sachs Harbour bury their dead. The graveyard lies on top of a hill behind the houses. One of the young women takes me there on her snowmobile. I cling to her back as she guns the motor, worried that we will hit something and keel over. The cold whips my cheeks. My lungs close up. But finally we reach the top of the hill, the sacred place. This is where the dead have been laid for decades, positioned to survey the expanse of Sachs Harbour as it grows with its houses and community buildings, the school, airport and the Arctic Ocean beyond. I stand there, rubbing my cheeks to keep the frostbite away, admiring the ancestors' view. I can read for the first time the shape of the land. Again I am struck by the absence of colour. Everything is white. The frantic yipping of the dogs bounces off the hills from far below.

Each grave is carefully marked. They are not square, but oval, surrounded with a circle of carefully chosen, rounded stones from scrub land. And some of them have sunk into the ground as the permafrost has vanished underneath the dead in the strangely warm climate. This is perhaps the most upsetting thing to the Inuvialuit of Sachs Harbour. Permafrost is a living thing to them. They count on it to make sure their ancestors have the best views. It is not supposed to vanish.

But at the feast, everyone is merry. I had expected musk ox or caribou or seal or fish. Instead, it is the really fancy food: turkey and hams with pineapple. Each little turkey cost about $50 (US).

Some of the women have made macaroni salads and fried bread. For dessert, there is jelly, poured while hot into Styrofoam cups and then chilled to perfection. The children are racing around in glee, aware that this is an important occasion.

And Sarah is here. She's been carried in and now sits in a wheelchair, bright eyes taking everything in. Before the meeting begins, before the chronicle of damage to their land is shown, she offers a prayer. Her voice fills the air, an urgent voice speaking in Inuvialuktun. The children, most of whom don't understand that language, are stone silent.

Sarah is giving thanks. For the food, the scientists, the meeting of the two cultures. Most urgently of all, though, she is praying for the earth.

AS I LEAVE to go back down south, Peter Esau hands me a package. It is a massive piece of Arctic char, frozen solid, from his shed full of food. He wants me to take it south to my children because they have never tasted Arctic food. He wants to send them a gift.

I travel for two days and two nights to get back home. Each night I set my parcel of char outside my hotel bedroom window, praying that the winter will be cold enough to keep the fish safe. I check on it several times a night. Then wrap it in the down coat to keep it cold in my suitcase during the day. By the time I get home, it is thawed. I cook it for my children. The three of us have never tasted anything so tender.

SIX

WHERE THE Rainforest GOES ON For Ever

Suriname, an obscure country in the Amazon rainforest, is suffused with equal parts fear and hope. At first, this seems paradoxical to me. But then I realize that each of these emotions, so primal, is the flip side of the other.

The fear is easy to find. Suriname, the former Dutch Guiana, is a society built on slavery. In the national museum in the capital, Paramaribo, ghastly iron collars studded with long, curved spikes and heavy, rust-pitted shackles are testament to the terror.

In gift shops, black-skinned dolls wearing traditional long dresses with a pillow-like hump sewn onto the back are on sale as curios for tourists. I ask about the significance of the hump and am told that during the nineteenth century the wives of slave-owners forced female slaves to wear this stifling costume as they worked in the sugar fields to make them less physically attractive to the wandering eyes and hands of plantation owners.

Outside the capital, the fear is more primeval. Suriname's dense tropical rainforest is home to the giant diamond-headed bushmaster snakes with fangs that can sever a human limb, to the fearsome jaguars and harpy eagles, top predators in the jungle. Famished piranhas course through every tiny tributary of the Amazon, incurable tropical diseases lurk in the saliva of legions of insects, and parasites and army ants are eager for human prey. And that's as well as the sheer physical extremes of pitiless heat, sopping humidity, typhoid-laced water and the odds of getting lost in the jungle.

The hope lies in precisely the same places. Slavery, hard fought and fearlessly defended, was eventually abolished. The modern Suriname is a rich and fantastical mixture of cultures, from Creole to Bush Negro to European to Javanese to Hindustani to indigenous people, all overlaid with the obsessive Dutch need for order. Even the national holiday – 1 July – celebrates the day slavery ended in 1863.

Most inspiring of all, the rainforest is almost entirely intact, part of the last tropical wilderness in the world. In fact, Suriname has the third highest percentage of healthy forest cover of any country in the world, after French Guiana next door and the tiny Solomon Islands of the Pacific Ocean. In the 1990s, cut-rate loggers from Asia were poised to begin cutting trees from about a fifth of Suriname, when conservationists stepped in and offered the Surinamese government a better deal to leave the trees standing. The trees stood.

At this stage, Suriname is the mirror opposite of Madagascar. Madagascar has less than 10 per cent of its forest cover left, and that is disappearing rapidly. Many of its unique, endemic species – including the primitive monkey-cousins, the lemurs – are at risk of disappearing too. As a result of all this, human hunger is widespread.

Here in Suriname, about 90 per cent of the forest cover remains, and it faces no looming threats. Much of the wildlife is healthy, including the monkeys. The people are prosperous. This tiny country on the northern coast of South America is one of the planet's few true conservation victories.

Much of that can be laid at the feet of Russ Mittermeier, the monkey scientist with the big, big personality who is president of the innovative Washington-based organization Conservation International. I have come here with him because it is his favourite place on earth, and I want to see what makes him tick.

I first heard about Mittermeier at the IUCN's World Conservation Congress in Jordan. Bored by the platitudes in a small session that ran before the conference, I turned to the IUCN's chief scientist Jeff McNeely, who was sitting beside me, and asked him who was on the cutting edge of conservation. He shrugged and replied, 'Russ Mittermeier', as if it were the most obvious answer in the world. So I figured that if I could come to understand Mittermeier, I would have a better sense of whether hope for the planet is warranted.

Just now, Mittermeier is on a quest. He has played host for a couple of days to some scientists from all over the world who have arrived in the tiny village of Kwamalasemutu, just above the equator in the south of Suriname. It is home to the ancient Trio tribe, whose population is now about three thousand in total, both here and further south into Brazil. The scientists are part of an urgent worldwide programme set up by the US National Institutes of Health to examine wild plants for medicinal properties before the species become extinct and their irreplaceable life-saving properties, evolved over hundreds of millions of years, are lost for ever.

And Mittermeier is interested in that, no question. It's just that some of the Trio men have told him that they have spotted a pair of harpy eagles and their young a few hours' paddle down the river. Since he got to Suriname, Mittermeier has been itching to see them for himself. He reminds me of a small boy who must wait to open his birthday presents. He's practically sitting on his hands. Now, finally, he's got the chance. We take to the canoe to search for harpies.

The Trio tribe are excellent at many things. They are superb hunters, artists and builders. But they are not excellent at making canoes. Trio canoes leak like the dickens. One of the Trio guides sits at the front of the canoe with his machete, chopping riotous vines and fallen trees so we can pass through. I keep wondering if he'll tip us over.

Naturally, we're not wearing life jackets here on this ancient tropical river. I look down at the muddy water, wondering what will happen if I fall in. I turn to Mittermeier and ask the question, mindful of every terrifying television special I've ever seen on the hazards of jungle life: are there piranhas in there?

He nods. Of course. This is a tributary of the Amazon. Lots of piranhas.

I'm covered from head to foot to protect against the malarial mosquitoes – boots, trousers, long-sleeved shirt with a collar, hat. Trying to be nonchalant, I wonder if these would weigh me down and make me easier piranha prey or protect me. I look at Mittermeier sitting next to me in the leaky canoe. He is in shorty shorts, no shoes or socks and his customary buff-coloured fishing vest with its manifold secret pockets, machete slung at his hip. He's ripped the nail off one of his big toes and is trying to dry it out in the blistering midday sun. He looks like Indiana Jones. This is a guy who has lived for years at a time in tropical jungles. He doesn't have a shred of fear. Just excitement because in all his thirty years in the jungle he's only seen two harpy eagles. He is certain of that because for thirty years he has written down everything he does in journals. He has two sets: one personal and another in which he catalogues scientific data. He times everything, writes everything down, collects superlatives – first, fastest, biggest – in every possible facet of his life. By this time, I've been with him enough to realize that this archive allows him to compete against himself as well as everyone else, against his own formidable previous bests.

HE EAGERLY SCANS the tops of the tallest trees as he tells me about the harpy, the mightiest, most remorseless predator in the South American skies. It makes the majestic bald eagle look like a crow. Harpies are a metre tall, sometimes more. Their legs are as wide as a woman's.

The harpy is endangered because so much of the rainforest in South America has been cut down, and they need wide swathes of forest in which to hunt. They usually eat monkeys or sloths, blasting through the forest canopy and ripping them straight out of the trees, but sometimes they go for children or small women, he says, eyeing me. Then he adds: 'Remember, they don't bother to soar. If you see it start to take flight *at all*, lie flat on your stomach on the bottom of the dugout and protect your head.' He demonstrates, wrapping his arms tightly around his neck and ears.

Suddenly, the Trio guides point to the upper branches of a kapok, the highest tree in the jungle. There's the harpy, guarding her eaglets, sharp eyes unblinking. Immobile, for the moment. I shriek and make for the bottom of the canoe, arms over as much of my spinal column as I can manage. Mittermeier whips out his binoculars. Chortles. Settles in to watch. This will only happen once in his lifetime and he knows it. 'That is really cool, guys. Really cool,' he says.

MITTERMEIER is an intelligent optimist, but a relentless one. He has to have hope. It's what keeps him going. He's been to eighty-eight countries – he's got two accordion extensions to his passport to prove it – watched the world's biological patrimony grow ever more endangered and spent a lifetime trying to save it.

He's not judgemental. He sees the dangers to the earth through the lens of biology, of Darwin's evolutionary theories. We're doing things the way the world taught us to do them thousands of years ago in order to survive. After all, he says, a lot of our behaviour is still hard-wired from when we lived in caves and painted ourselves blue. And much of it is worth celebrating. All in all, though, he has a hunch that the people who live in places like these jungles of Suriname – he gestures to the wild rainforest on either side of the Trio canoe we are in – will have a better chance of surviving over time, especially surviving the diseases that are bound to attack the masses of human protoplasm on the planet today.

This surprises me. I would have predicted that humans in the centres of modern civilization would have a better chance of survival against any disease. I think of the billions of dollars spent on medical equipment and training and drugs in the developed and developing worlds.

Mittermeier sees it a different way. He looks not at the knowledge of a species such as *Homo sapiens*, but at the biological system as a whole. Whenever a species becomes abundant, it becomes prey. Right now, there's a ton of human protoplasm out there and not many new predators. But parasites and viruses can evolve in a heartbeat and jump from host to host, species to species. Look at AIDS, sweeping through Africa ahead of all attempts to vaccinate against it. And old diseases are coming back newer and tougher: look at cholera and malaria. Humans are just a meat market for predators like that. And we're all handily bunched together by the millions in big cities, and prone to travel quickly by aeroplane from one end of the planet to the other.

He cuts off excitedly and points off to one side of the canoe. A screaming piha, a drab brown bird that makes a gigantic lek, or mating ground, on top of the forest canopy. And over there, one of the yellow-headed vultures.

He pauses before launching back into his theory. All the biological signs would say that humans are at considerable risk of an outbreak of disease that we have no idea how to control. Africa and India could really be in trouble. Some places are going to be hammered, he says. It feels like a certainty.

That's why it's such a relief to him to come to Suriname. He first arrived here in 1975, a precocious monkey researcher in his twenties, anxious to find the perfect tropical rainforest to do field studies for his doctorate. He found it in Suriname's untouched interior and proceeded to live here for the better part of three years, becoming the first to map the behaviour of all eight species of indigenous monkeys. He fell in love with everything about it: the forests, the monkeys, the people. Now, he comes here to find respite from the meetings, the phone calls, the pavement. It's one of the few places in the world that hasn't changed in the decades he's worked as a conservation scientist. It's where his hope lives.

And it's not just the animals and plants. While so many other South American tribes have become culturally or biologically extinct, especially those along the Amazon River, the Trio have survived. Yes, the young generation may wear T-shirts with American pop stars on them and plastic flip-flops on their feet. The men may wear jeans and button-up shirts. But many still have the full black fringe, long black hair and loincloths of generations past. Their legends are ancient, rituals rich, lifestyles similar to that of their forebears. I think of the Yamana of Tierra del Fuego, at the bottom tip of this continent. Darwin mocked them and his captain tried to civilize them. It was disastrous, and the Yamana died out. The Yanomamo not very far south of here are celebrated internationally in song and on stage simply for continuing to exist.

THE TRIO HAVE had a lot less attention. But that may change. Mittermeier has been negotiating for months to see one of the most sacred of all Trio cultural artefacts: the legendary caves of Werehpai. He thinks they may make a superb offering for carefully controlled tourism. That's part of Mittermeier's philosophy, no matter which country he is working in: you can't focus only on the forests and the animals, you have to focus on the local people too, and on what they want and need. The collective will to leave the forests standing must come from them.

The Trio have passed down an intricate tale about the caves and their cannibal inhabitants for thousands of years. But despite the descriptions, the caves themselves could never be found – until a few months ago when a Trio hunter from Kwamalasemutu stumbled upon them. Mittermeier was one of the first non-Trio to be told of the discovery. He is known and trusted here, but this particular part of the Trio legacy is considered too sacred to show even to him. Last time Mittermeier was here, the tribe promised him a glimpse, but backed down at the last moment. Only one non-Trio has seen the sacred caves, a photographer invited by the Trio because he had the equipment to take pictures.

This time, though, they have promised Mittermeier that they will show him the caves. He's keeping his fingers crossed, aware that they might change their minds at any moment. So we're tramping with a guard of about a dozen armed tribesmen, plus the chief, through a magnificent stretch of dense tropical rainforest deep in the jungle on our way to the caves. Underneath is Precambrian rock from the oldest of the earth's geological eras, when organic life, if there was any, had no hard skeletons. Mittermeier's eyes are focused on the ground, casually watching for poisonous snakes, ferocious army ants and baseball-sized tarantulas that he won't mention unless he has to. Hundreds of species of trees arch high above. Some are so old, it would be hard to encircle them with human arms. They thrive despite the thin tropical soil, held up by weird support contraptions that grow out of their bases and remind me of the flying buttresses of Gothic cathedrals. Verdant palm trees grow everywhere, glossy in the dappled light, one of the key signs of a healthy forest understorey. Liana vines loop down from the treetops, insinuating themselves into the branches of the trees. The forest is so wet, so hot that, to my northern nose, it smells like stewed rhubarb.

WE COME TO A CLEARING that the Trio have made. Already, this is a sacred place because it is so near the caves. The chief launches into a long speech and something that is clearly a prayer. The few non-Trio among us are holding our breath, waiting to see if we will be allowed to enter the caves. The sense of mystery is palpable. The Trio look strangely implacable, as if they have reached some sort of tough decision. These are the caves made sacred by the ritual spilling of human blood and the eating of human flesh. They are alive to the Trio. And here we are, the first non-Trio – apart from the useful photographer – to be led to the caves. I wonder if the caves are longing to be sanctified again. I begin to smell fear, and I realize that I am drenched in the sweat of terror.

The Trio order us to take off our cameras. Then, armed with machetes and rifles, they frisk us and search our bumbags. We must leave our backpacks behind. There is to be no record of this. Then they allow each of us to go forward the final several hundred metres to the caves. But they deliberately separate us, blocking us from each other with the bulk of their bodies. I am intensely aware of the fact that we are outnumbered two to one, and that the Trio are in superb physical shape. All of us are terribly alert. I can taste the salt on my lips.

And then, the caves of Werehpai. There are probably fewer than a dozen, connected in a bewildering maze. Some are large enough to hold a dozen people. Others are so low we have to crawl on our bellies. Every surface of rock is carved with large crude figures: here's the face of a paramount chief, with his elaborate headdress adorned with the spiky outline of harpy feathers. There's a bushmaster snake, poised to strike. That one looks like a human figure with stomach excised. Just beyond is a broad, flat stone the circumference and height of a table. It has the feel of a sacrificial altar. I stumble and touch it to steady myself. My guards stiffen and the other guards swirl fiercely around. Mittermeier tells me urgently to take my hand away. Then I notice the dominant carving, repeated everywhere. It is a single, wrenching face, eyes wide open, mouth stretched in terror.

THE TRIO CAN READ the legend of Werehpai on these walls. She was a member of the Akijos, a cannibal tribe said to have lived in these caves thousands of years ago. The Akijos were much more technologically advanced than other tribes of the day. They could paint, draw and weave. But they used their advanced knowledge not for good, but to terrorize other tribes, taking their people to satisfy grisly cannibalistic rites. One day, they stole two young children – a girl and her brother – from the Samuwakan village further south and gave them to Werehpai to raise. When the girl came to a certain age, the Akijos tattooed her and then ate her alive, keeping her conscious for two days while they consumed her, as the ritual demanded.

Werehpai knew that her people would also do that to the boy, Aturai. So she warned him. She told him that his own Samuwakan village was near a big stone mountain. If he got to a high place, he would be able to find it. So Aturai, who was a good hunter, told the Akijos he was going shooting with his bow and arrow. He shot at a toucan and deliberately missed. And as he went to look for his arrow, he ran off.

Aturai found his way home, and his people were happy to see him. But when he told them what had happened to his sister, they wanted revenge. So he took the skills of the Akijos, made breastplates of bamboo and mud and trained an army of warriors. Then he brought them here to these caves and killed all the Akijos, including Werehpai herself, who had saved his life.

The people of Kwamalasemutu believe that the cave carvings are the record of that battle. As I look around me at these vivid petroglyphs, made perhaps five thousand years ago, I almost choke, imagining the smell of the burning torches that people once must have held to read this macabre history book. I can see the shifting shadows, hear the shrieks of agony, smell the blood.

It has the feel of a place abandoned in a hurry. There's a scattering of axe heads, not fully formed, as if the carvers were interrupted in the process of making them. As we crawl through some of the lowest caves, we see intact pots and bowls made of red clay. Kamainja Panashekung, the Trio hunter who found the caves, shows Mittermeier the final prize. Behind a rock plate in what looks like a secret compartment are two prehistoric carving tools, likely the ones artists used to carve this rock. Mittermeier shakes his head in awe. They probably haven't been touched for thousands of years, he says. But they look like they were left there just yesterday.

As we round a corner into another part of the maze of caves, there's another rush of adrenalin. A jaguar, the deadliest land predator in South America, has made a home here in the caves. Bones of a wild pig it has recently caught for lunch are laid out on another rock, broken and sucked clean of marrow. Fresh footprints are visible in the cave's sandy bottom. The smell of the cat is acrid, overpowering. I ask Mittermeier if it could still be in the cave, hiding in a shadow, waiting to pounce. Maybe, he answers. But humans are only *occasionally* jaguar prey, he adds, trying to comfort. We're a little too tall for all but the most desperate jaguar. I keep looking around in the dark, searching for narrowed yellow eyes, wondering what would make a jaguar that desperate.

At last, we wrench ourselves out of the caves and go back into the flickering sunshine of the forest. I feel spent and calm, marvellously serene.

THE TRIO, TOO, WONDER whether the caves and the lush rainforest might be good tourist fare. They have been preparing a spot where people could stay if they made it this far into the rainforest – a couple of hours by Nomad charter plane to Kwamalasemutu and then a bit of paddling down the river after that. It's a meticulously cleared spot with some covered shelters where people can hang hammocks. That's the only safe way to sleep in this part of the jungle, off the ground and swathed in mosquito netting. Mittermeier has been telling me that when I wake up in the morning, I should be prepared for the fact that the top of the netting, right above my face, will be sagging with tarantulas that have been prowling around in the night. I lug my backpack into the back of the shelter where I'll be sleeping, change into something not drenched in sweat and suddenly feel a searing pain on my butt. Mittermeier rushes in at my scream. 'Is there unbearable pain spreading through your body?' he asks.

'No,' I reply.

'OK, then,' he says cheerily. It's probably not a deadly snake bite. He sends in a Trio healer, who nonchalantly examines me in the half-light and pronounces his verdict. It's the bite of a leaf-cutter ant. The whole camp explodes with laughter.

Mittermeier announces that he is going for a dip in the river to cool off. My jaw drops. Are there piranhas in there? I ask.

Of course, he says, shrugging. But they usually go for you only if you're bleeding.

I think about the toenail he's ripped off. He's been dousing it regularly with straight hydrogen peroxide to help it heal, an old trick he once learned from a jungle doctor.

What about your oozing toenail? I ask.

Nah, he says. Probably the worst they'll do is come over to check me out. You can feel them when they do that: Ping, ping, ping against your leg.

Some people are terrified of snakes. Some of spiders. For me, the primeval fear is piranhas. I put on my bathing suit but just sit on a rock, watching Mittermeier splash about. I go so far as to dip my hands in the water and wet my sweaty face. Finally, I muster the courage to swing my legs briefly in the water. It's the most I can do.

LATER, WE GO UP THE RIVER a little way to visit another small Trio settlement. Mittermeier needs to do this, to clap eyes on every single member of this tribe he can. In part that's because he's been coming here so long he knows many of them, has watched them grow up and grow old. In part it's because he is an obsessive collector of Surinamese cultural artefacts. He's already made a voracious tour through Kwamalasemutu, buying up every set of bow and arrows, every beautiful seed belt they will sell. They make things for him in between his trips. But this little village might have something else, and he can't pass up the chance.

He's right. One of the tasks that falls to the women is to grate the starchy root of the cassava, or manioc as it's called in other parts of the world. To make their graters, they collect thousands of tiny, sharp stones from the river, each perhaps the size of a grain of rice. Then they pound them into a piece of rough-hewn wood. Sometimes they decorate the finished product with the blood-red juice of a local berry and then drip thick wax on top. The women are using several of the graters right now as they prepare supper. To Mittermeier, this is the best possible type of artefact: one that is actually in use. The women trot out all the cassava graters in the camp, every one streaked with the white residue of pulverized cassava. Mittermeier wants the biggest, which is also the most cherished in the camp. He offers them a huge sum for it, by Trio standards. The women resist for a while, realizing that they will have to go through the laborious task of making another if they sell this one. Finally, they relent. Mittermeier crows. He has a museum-quality collection of Surinamese artefacts back in Washington, gathered like this over the decades. He figures most of them would have rotted in the jungle humidity long before now and at least he's preserving some of them and passing money to locals. But he also knows that he simply loves to collect. It's a need he can't fully explain, but it's as much a part of him as the need to conserve the biological riches of the planet.

We're just ready to leave, when one of the camp's hunters arrives with fresh meat for supper. It's a capuchin monkey he's just shot. Its tiny fists and feet are clenched in the spasm of recent death. The hunter's wife starts shaving its black hair off with a small machete, preparing it for the meal. It's the same nonchalance I've seen when fishermen casually gut a fish. Without the hair, the monkey's white skin is the colour and texture of a child's.

As we go back to our own camp, Mittermeier shakes his head. Ecotourists would love to see the caves of Werehpai, maybe the harpy eagle. And it would be a great incentive for the Surinamese government to keep the forests standing, and to support the ancient lifestyle of the Trio. But this monkey-skinning part just wouldn't be a hit. Tourists would rather see monkeys in the trees than on a spit.

FURTHER INTO THE country's interior is the heart and soul of Mittermeier's hope for the planet: the Central Suriname Nature Reserve, 1.6 million hectares of protected Amazonian rainforest, the single most pristine tropical forest on earth. It is a living repository of a great deal of the biological diversity now on the planet and part of the timeless womb of future creation. Watersheds flow unrestricted here; jaguars roam unmolested; the assembly of life is as complete now as it was when Darwin wandered in the South American forests in the nineteenth century, or when the Akijos feasted on human flesh five thousand years ago.

In 1997 Malaysian and Indonesian loggers set their sights on Suriname, bidding to cut down 3 million hectares, or about a fifth of the country. The Surinamese government was close to approving the plan. The people of Suriname had no problems with it. But Mittermeier did, and that's when he came up with the idea of the massive reserve.

He worked closely with Willem Udenhout, who was Suriname's ambassador to the United States and Canada for part of the 1990s and special adviser to President Jules Albert Wijdenbosch. During this period, Udenhout attended a board meeting in Washington of Mittermeier's organization, Conservation International. He knew of the group from his years as ambassador to the US and recognized that its leaders were too savvy to expect the Surinamese to forgo logging income in the absence of something else.

'Environmentalists who are innocent of economics have no audience,' he told me when I saw him in Paramaribo.

The board meeting was one of the rare, magical moments of Udenhout's life. There was Mittermeier, dreaming aloud about joining up three existing protected areas in Suriname to create an expansive new reserve. And in the process, he was steering this New World country into a pattern utterly unlike those of its neighbours or most of the rest of the world: treating standing forests as a resource instead of the lumber from trees that are cut.

'When he gives his imagination free rein, it's like you're entering his interior monologue and you just happen to be there,' said Mr Udenhout. 'After he finishes talking, he sees you are there and says: "You think we can do it?"'

In the end, the plan was so successful that Udenhout says preserving the forests has become part of the overall economic plan of the Surinamese. The reserve now includes one of the largest protected tropical river systems in the world and has been designated a World Heritage Site by UNESCO. It's supported by a fund of $15 million (US) that Mittermeier and CI put together. Eventually, the reserve could gain Suriname some credits under an international carbon-trading system being set up under the Kyoto Protocol to reduce greenhouse gas emissions. That would make it even more lucrative for the Surinamese.

The government has issued new currency reflecting

this shift in values. Before, the pictures adorning the bills were of traditional New World pursuits such as logging, oil extraction and mining. They've been replaced by pictures that celebrate wildlife and forests.

As for the ambassador, two years ago, after the 1.6-million hectare Central Suriname Nature Reserve was created, he signed on to work with CI as its executive director for Suriname.

THAT'S ONE OF THE POLITICAL levels Mittermeier and CI work at. But Mittermeier has long believed that it's important to understand the people living in the areas and their local political leaders too. You can't just impose a protected area from on high. That's why we are going to Asidonhopo, the seat of power of the Maroons, more correctly known as the Bush Negroes. These are the extraordinary people whose ancestors not only survived being brought to Suriname from Africa by slave traders, and then the infamous brutalities of Dutch slave masters, but also the rigours of the jungle after they escaped.

Mittermeier has a secret theory that the Bush Negroes have undergone the most rigorous natural selection of any group of *Homo sapiens* on the planet. Evolution has pared this gene pool down to one that can withstand extreme heat, humidity, swamp diseases, can outwit piranhas, crocodiles, jaguars and harpies. The first Bush Negro I see after we land in our tiny charter plane is a young woman who looks about nine months' pregnant, heartily paddling her canoe. The people are short, and so muscled that they look rubbery. They wear no shoes as they walk through the jungle, but they have thick, leather-like calluses on their feet. One of our guides has no toenails, just tough skin where they once were. The women are among the very few people on earth that Mittermeier is wary of. One came up to him a few visits ago and ordered, 'You do me.' He blushed and said his wife wouldn't like that. She replied, 'She's not here.' He had to wriggle his way out of the encounter.

Perhaps three hundred people live in Asidonhopo, including Songo Aboikoni, the Granman, or paramount chief of the twenty thousand or so people of the Saamaka tribe spread over sixty-one villages. The chief's name means 'the boy is clever' in his language, and he's renowned for his tactical skill. Mittermeier wants to check in with him and see how things are going.

So the first thing we do when we arrive at Asidonhopo is go to a ritual *krutu*, or audience with the Granman, at his residence. It's not done to visit this village without telling him why you are here.

As Saamakan custom dictates, guests must leave their shoes outside before they enter the chief's house. The Granman is barefoot too, dressed in a crisp grey tropical suit. There is a picture of Mittermeier on the wall behind his head, in the place of honour. He does not speak directly to us, not even to Mittermeier. He speaks to a *bassia*, sort of a deputy minister, who then communicates for him.

The message is harsh. The Granman is not happy with Mittermeier, who the Bush Negro have known for decades. Haven't the Saamaka always been his friends? And now there's this great new Central Suriname Nature Reserve just beyond Saamaka traditional lands, and a fat trust fund of $15 million to support it.

The Saamaka, though, don't see a penny of that money. Their buildings are crumbling. Their young men are going next door to French Guiana to earn money. The central government of Suriname gives them nothing. Nothing. What will Mittermeier do for his friends the Saamaka?

The CI contingent in the room shifts uneasily. This is not what they expected. They don't have a ready answer.

ONCE OUTSIDE AGAIN, though, there's a welcome wagon for Mittermeier. The people here know that he's an inveterate collector of Surinamese artefacts and that they can score what is to them big money. In fact, he will buy almost anything they want to sell.

So here come the villagers with paddles, intricately carved, that they use in their exquisite canoes (which don't leak, unlike those built by the Trio). Mittermeier is in heaven. A CI worker walks alongside him, a big blue duffle bag over his shoulder. An hour later, it's stuffed with items the Saamaka use every day. The people have also come out with things they consider junk. Mittermeier loves it. He picks up an antique stool. The patterns, colours and shapes might be from the Ghana of a hundred years ago. Some of the older women offering these goods for sale have the ancient marks of African scarification on their faces and bare breasts.

Mittermeier is a field scientist, and so he has a system for such purchases. Each time he buys something, he makes a note in his tiny coil-bound book of the person who made the item, the date he bought it and how much he paid. He writes the same information in black indelible ink on a museum-style tag and attaches it to the artefact. Then he gets the maker to pose for a photograph, holding the item he's bought. Finally, he takes a Polaroid of the same scene and hands it out. It's one of the ways he keeps track of the Saamaka families. Maybe more important still, it's a way for the Saamaka to remember him and CI. Every time he shows up at Asidonhopo or another Bush Negro village, faded Polaroids with his markings are waved jubilantly in front of his face.

SURINAME'S HISTORY of slavery is never far from the surface, no matter what part of the country you're in.

But here in the domain of the Bush Negroes, the whole culture and location of their traditional lands is based on the need to escape slavery.

The Dutch were renowned for their cruelty. About 330,000 African slaves were brought here, approximately the same number that arrived in the United States. By the time of emancipation, the United States had 1 million slaves. Dutch Guiana had only 25,000 left, Mittermeier tells me. The Dutch were especially odious to slaves who escaped and were hunted down and caught. They hung them on meat hooks in a square in Paramaribo and hacked them apart, limb from limb. Sometimes they just cut the breasts off the women.

The story of slavery is told in dance too. Late that night, as the heat slowly dissipates, we go to a performance of the *seketi*, a traditional Bush Negro dance, at the village of Pende Konde, just up the river from Asidonhopo.

About a dozen women seem to appear out of nowhere, dressed in knee-length, brilliantly coloured cloths tied around their waists. Up on top, they wear skimpy vests or bikini tops, their dark skin shiny under a makeshift light. It's a long way from the humps and sweat-making dresses their ancestors were forced to wear on the sugar plantations.

They bend at the waist, looking down, as if they are cutting sugar cane, and begin clapping a rhythm as old as time. Their voices rise, carrying the beat. The song is based on those their forebears sang as they worked the plantation. Forbidden to talk to each other, they sent secret messages encoded in their songs. They would sing, for example, of a bird that flew to the left of the house, a signal that an escape would take place there later that evening.

The *seketi* is still used to voice messages too painful to be said directly. One of the women, skin glistening, hair in neat cornrows, has a message for her husband. Her voice is slow and mournful against the impassioned rhythm: you keep apologizing, she sings, but now I realize you don't know how to take care of a woman who loves you, so I'm leaving.

Mittermeier has been snapping photographs. He stops, realizing that this powerful message out of the blue has a special meaning for me. I've been telling him about the pain of my broken marriage.

At the end, I ask whether he arranged for that song to be sung. He grins wolfishly, loath to relinquish any shred of potential praise, trying to figure out how he can take credit for it. Finally he confesses that it was just a fluke. Women in that part of the world have to leave their husbands too.

In the morning, before Mittermeier has even had his breakfast, a handwritten letter comes from Granman Songo. It's an offer to add as much as 400,000 hectares of Bush Negro traditional lands to the Central Suriname Nature Reserve. The land would include the critical headwaters of the Suriname River. This is one dream Mittermeier has never put words to. He is flabbergasted.

'That's pretty cool,' is all he says.

THE SCARY THING is that the business of species conservation is sometimes just that fragile. Sometimes, it's just a question of happening to be in the right place at the right time. Miss the plane, cancel the trip, get a touch of malaria and, bam, a species goes extinct or gets that much closer to the brink. Mittermeier has been telling me about the Bali mynah. It's a brilliant white bird with a blue eye that lives only on Bali. There are just a handful left in the wild because the birds have been caught so efficiently for the trade in live birds.

But they breed well in captivity, and there's a programme at Bali Barat National Park to increase their number. For some reason, though, the programme wasn't working well. Mittermeier was in Bali and happened to have some time to check it out. It turned out the wardens who were supposed to stand guard over these mynah birds were so poorly paid that they didn't have enough money to buy petrol. That meant they couldn't keep up with the armed poachers who kept invading the mynah-breeding centre to catch the birds. A cheque for $6,000 (US) was all that was needed to solve the petrol problem, although the programme is still far from successful.

CI was born out of luck, timing and detemination, too. It came to be in a single weekend in January 1987 out of a blaze of dissatisfaction felt by about fifty scientists and conservation workers who were with other organizations. The CI founders were convinced that countries can build an economy around conservation principles. In other words, they don't believe that a rich piece of landscape must remain utterly untouched and its inhabitants supported by money from elsewhere. They also don't believe in doing good just for the sake of it. There has to be something to measure at the end of it. It was a heretical idea at the time and is still disparaged by some environmentalists. The World Wildlife Fund US, some of whose members defected to CI, gave the members a cake-mascot to mark the founding of CI. It was a vulture.

Mittermeier has a quiet disdain for people who just march in the streets and write thoughtful tomes on species conservation. The real way to save endangered species – and therefore biological systems that also serve humans – is to be there on the ground finding out what's happening and raising the money that can solve so many ills.

That's how CI works. It is one of the most successful fundraising organizations in the United States. CI is about money as much as it is about conservation. It raises in the order of $50 million (US) a year from blue-chip corporations, foundations and wealthy individuals, including some of Hollywood's leading lights. The actor Harrison Ford is a passionate board member and so is Gordon Moore, a co-founder of Intel, who recently gave CI a gift of $261 million over ten years.

ALTHOUGH HE ADORES every hectare of Suriname, Mittermeier is more in love with the area around Raleigh Falls in the Central Suriname Nature Reserve than he is with any other part of the country. His research station from the 1970s is just down the river. Some of his botanical marks are still on the trees.

Right now, he's on a mission to see some of the eight species of monkeys that live in Suriname, despite the fact that his big toe with its ravaged nail is incredibly sore. He winces with every step. That's one of the worries in the jungle: it's so humid that sores sometimes don't heal. It's like the clothes. You get something wet and it may not dry for weeks.

The jungle is pristine, teeming with life forms most people have never heard of: the great tinamou bird that lives on the forest floor and lays a blue egg, the cock of the rock with its brilliant red plumage, the forest floor toad with its pointy nose that adapts its colours to whatever part of the forest it's in.

Mittermeier is watching the ground, as usual, for dangers. He knows this forest so well that he will point over to a waterfall and tell you the story about the guy he was with once who got bitten by an electric eel over there and almost died. But he's sniffing the air too. He can tell which monkey is above him in the trees just by the scent it gives off. The big red howler monkeys are especially smelly. This is his idea of heaven. All the way from here in the middle of Suriname south to the border of Brazil, the rainforest is uninhabited, pristine. To Mittermeier, this vast swathe of forest that he helped save is just a small piece of a massive global puzzle he is working on. He's already thinking about how to conserve forests in the rest of the Amazon, the Congo Basin, New Guinea. He's not happy unless he can see the big picture.

I'm thinking about Darwin again. About the fact that human ancestors evolved in jungles like this. Surviving in a place like this honed our complex ingenuity but also limited our ability to see the big picture. I've been reading *New World, New Mind* by Robert Ornstein and Paul Ehrlich, about how the human mind evolved.

Ornstein and Ehrlich argue that, historically, the human brain has had to limit the amount of information it takes in. Taking in too much causes confusion. Instead, the human brain is geared towards reducing complexity. That's what allowed our ancestors to react quickly enough to survive.

Not only that, but because *Homo sapiens* is a rather new species and because evolutionary change in our species is slow (as opposed to the rapid modern change in, say, the AIDS virus), our brains are little different from when we were hunters and gatherers fifteen thousand years ago, living in caves, like the Akijos.

Other traits fostered by human evolution include translating information into the immediate and the personal. That means the human brain has a bias towards new news, as opposed to long-term trends. It's why, for example, pictures of specific big-bellied children starving in an African country can successfully raise money, but lectures on the geopolitical origins of poverty cannot.

Ornstein and Ehrlich believe that perceptions of gradual, long-term trends were actively suppressed as humans evolved because individuals didn't have the ability to do anything about those trends. In other words, the early humans who spent their time worrying about whether the jungle would be there in three generations were snuffed out in favour of the ones who could find food in it today.

Are humans incapable of seeing the big picture, I wonder. People like Mittermeier, Leakey, E. O. Wilson, Rachel Carson, like Ehrlich and Ornstein, like Myers and Dallmeier and Sarah Kuptana and Darwin know what's going on. It's just that they are not the norm. As a species, it's not our strong suit to think long-term.

I remember Cassandra, the cursed princess of Virgil's *Aeneid*. Daughter of King Priam of Troy, she was fated to tell the truth and never be believed. The night before Troy fell, she alone knew that the Trojan Horse would be its undoing.

She cried out her prophecy. Everyone ignored her. The next day, the Greeks, lodged in the horse's belly, broke out and turned their fury on their Trojan foes. Cassandra's fate was to be raped and then to commit suicide as she was handed over to yet another Greek leader for violation.

But those too blind to believe her were cursed too. Unwilling to be inconvenienced by the truth, mired in denial, they told Cassandra she talked too much. Troy fell. The Trojans were slaughtered.

I'M NOT SPENDING enough time figuring out how to get through today either. I jump off a fallen tree onto the path in front of me to follow Mittermeier and land first on one side of my foot and then the other. It starts to swell alarmingly. Mittermeier comes back, examines it, and says it's probably just a sprain, not a break. But as I continue walking through the day, it gets bigger and bigger. Back at the camp, two of our guides take me in hand. One wanders out into the jungle in the dusk and returns with a huge handful of herbs. He puts them in a basin of water and gently lowers my massive ankle into it. Half an hour later, another of our guides cradles my foot on his lap, presses two fingers to my ankle, shuts his eyes and concentrates hard. He tells me he's transferring energy from him to me to heal the hurt.

Within the hour, the swelling has abated. It's still sore, but I can walk on it.

Good thing, because the pièce de résistance for Mittermeier is yet to come, and there's no way I'm going to miss it. The whole trip, he's been longing to climb the Voltzberg Dome, a piece of sheer Guyana shield poking hundreds of metres up through the jungle floor at a 45-degree angle on its gentlest face, a vantage point to survey the Central Suriname Nature Reserve.

He once climbed it in eleven minutes, but that was 'some macho thing', he confesses. Usually, it takes about twenty minutes if you really climb hard.

He's only been up here once since the big reserve was created, and he needs to go up again. It's brutally hot. Wild pineapple and vanilla cling to the bare rock. The Bush Negro guides climb in bare feet.

Mittermeier slows down to help me because my ankle is still sore, nobly relinquishing his place at the front of the troop and letting others reach the summit first. It's a huge sacrifice. Still, he notes, looking at his stopwatch, we made it in nineteen minutes, which is fairly good. But those other guys made it in sixteen to seventeen minutes.

He looks around. He is the tallest creature about, on the highest surface in sight. He can see to Brazil, to Guyana, to French Guiana. And as far as the eye can see, right to the horizon in a 360-degree radius, the rainforest is pristine and more or less uninhabited by humans. It's one of the rarest sights on the planet. Mittermeier waves across a vast swathe. All this would have been logged for a pittance if the government hadn't agreed to make this reserve.

Instead, the macaws fly overhead, their colours brilliant against the clear sky. The jungle below is astonishingly uneven and complex. No two trees seem to be the same height. No two are the same shade of green.

Deep under the rainforest canopy, a howler monkey lets loose with a roar. It sounds exactly like a lion bellowing into a microphone. The vibrations seem to bounce through the forest, the way they must have done thousands of years ago.

Mittermeier could be forgiven for basking. But he doesn't. He's already trying to figure out how to save the next bit of the planet. If it can happen in Suriname, it can happen in other places, he figures.

He thinks of his son John, the oldest of his three children. At fourteen, John is already a gifted ornithologist who has spotted birds in twenty-five countries. One of the great highlights of his bird-seeking life happened the previous summer, when Mittermeier took him to Brazil to see the last Spix macaw remaining in the wild. It's a graceful bird, light blue with a long tail. One of the true glories of nature. In November it died, leaving only a few live specimens in zoos. John, who attends a boarding school in the United States, went on a one-day hunger strike.

Mittermeier, sitting up here on the top of the Voltzberg in the punishing heat, gets worked up just thinking about it. He called John.

'I said, "Cut the crap with the hunger strike."'

Instead, he told his son, figure out how to breed the bird in captivity and reintroduce it to the wild. Or go save something else from going extinct. John began to eat again.

IT'S ONLY ON OUR WAY BACK that I realize what my fear of the piranhas means. Like all powerful phobias, it is strongest when imagined and never faced. The piranha is a symbol of the frightening things I have been writing about and travelling around the world to see: mass extinction, climate change, Darwin's pathological reluctance to rock the theological matrix of his day, our society's reluctance to admit that the planet's resources are exhaustible. And it's a symbol of my great personal terror of leaving my husband to make my own life.

All of it, every bit, has to do with fear. But on the other side of fear is hope. Maybe faith. So the final time I ask Mittermeier whether our potential bathing water has piranhas in it, he snorts a bit and sets out to show me.

He's sitting on the deck of a fabulous new cabin built at Raleigh Falls to draw tourists to Suriname's wilds. Below is a sandy beach on the river's shore. He tosses a piece of cooked chicken from the meal we've been enjoying into the shallows.

It takes a minute, but the first piranha – no bigger than a finger, with a neon-red tail – finds the treasure. Almost instantly, a swarm of the tiny fish appear, voracious and swift. For a few seconds, all you can see are bubbles, rising to the surface as the fish become more and more frenzied.

Mittermeier provides the soundtrack: 'Bam. Bam. Bam.'

By the time the water clears, the chicken has vanished. Mittermeier chuckles and walks away. 'Cute,' he says. I am silent in shock. Of course, these are the tiny piranhas. The big ones, also damnably numerous in this river, are the size of a frisbee. Their teeth are like a big dog's. Except sharper. We can tell because one of our travelling companions – Daniel Johnson, the American ambassador to Suriname and a committed fisherman – has just caught three of the big ones and is flexing their jaws, showing us.

It's clear to me that unless I swim with the piranhas, I will be not only consumed by fear but also untouched by the hope I seek. I will be unable to believe that humans, who I know have given up even such ingrained practices as slavery and cannibalism, will also give up the fable that they can keep harming the earth.

So I put on my bathing suit and descend to the river

to swim. It is a blessed relief from the sweat and heat. I feel a few piranha pings, but nothing to worry about. I emerge later, triumphant, whole and clean.

For supper, we cook the ambassador's big piranhas. They are full of bones and sweet, sweet meat.

SEVEN

Iceland's New Power

Iceland's secret is that it is not a land of inert ice but of kinetic steam. It is one of the few places in the world in the throes of constant geological creation, driven by the force of heat far below the surface. Volcanoes spew lava, and geysers boil water on this metamorphic island. Earthquakes daily crack open new fissures in the land, exposing the roiling inner workings of the planet. In ancient times, the island was thought to contain the maw of hell, a voracious opening to the eternal fires of the netherworld.

Things here are primal and, cajoled by the steam, always changing. An island will pop up off the Icelandic coast and then gradually wear away in front of the eyes of geologists studying it. The very size of Iceland – 103,000 square kilometres – is understood to be changeable, shifting in response to powers deep within the earth.

THE QUARTER-MILLION people who inhabit this strange land revel in the fact that they, too, are adept at metamorphosis. They believe in the power of transformation and also in resisting transformation if necessary. They are joyfully, avowedly, proudly odd. They shift roads to avoid disturbing the lairs of elven folk who live beneath the ground. They embrace modern technology if it suits them – mobile phones are more common here than anywhere else on earth – but have refused to alter the Icelandic language since the Middle Ages. The language here is so proudly unadulterated by foreign terms and English slang that Icelanders make a point of coming up with new, purely Icelandic words for computers and other modern technology.

Underpinning all this social and geological strangeness is a rich set of cultural legends cast with extraordinary, complex characters whose deep psychological wisdom has been passed down over countless generations in the form of sagas. These sagas have the potency of holy stories for Icelanders, set down in medieval manuscripts that describe in bloodthirsty detail the epic tales of about forty families.

While each of the adventures is different, the central theme is common: Icelanders are born to struggle but triumph. These are legends that demand evolution, innovation and vision. They put a premium on exploration, on cleverly making use of what's there. In fact, to an outsider, it feels as if being an Icelander is the equivalent to having an obsession to adapt.

THE STEAM HAS drawn me here. And the spirit of innovation. Iceland is the first country in the world to announce its intention of doing away with fossil fuels in favour of harnessing the mythical energy of hydrogen, which it can do with steam. Making use of hydrogen is an idea that has kept much of the scientific world enthralled for decades and is widely seen as an impossible dream. But a compelling one. If someone could figure out how to capture the thermal energy contained within hydrogen and convert it on the spot to electricity capable of being stored and transported, it would change who controlled energy – and therefore wealth – around the world.

Right now, the countries where people do well economically have access to energy, and that's all based on extracting or harnessing substances from the earth: wood, running water, coal, gas, oil. The problem for countries that don't have those natural resources is that they have to have them brought in. And in an isolated, resource-poor country like Madagascar, the distribution system for gas and oil is mostly absent and would be costly to expand, especially when the populace is far too poor to pay for the products in the first place. It's a superb Catch-22.

The current sources of energy affect the global environment as well as the economic well-being of humanity. Vast tracts of forests are being cut down for timber and paper, but also to burn for fuel. Waterways are being harnessed and re-routed, disrupting the planet's hydrological systems on a wide scale. Oil and gas are being burned for energy at increasing rates across the globe, sending so much carbon dioxide into the atmosphere that it is destabilizing the global climate. Many countries with the most important stores of the planet's biodiversity are also the most poverty-stricken and therefore the least able to take care of the environment. If the sixth extinction spasm is to be halted, those countries will have to have enough wealth to put conservation at the top of their list of priorities.

In other words, energy, poverty, environment, climate, wealth, biodiversity and extinction are all connected in an intricate web. Touch one part and the whole web shivers. If the web could be unwoven so that energy affected wealth but not the health of the planet, it would mean a massive advance for humanity and its shelf life.

That's where hydrogen comes in. It is the most common element in the universe. And it is protean: it is one of the two components of water (along with oxygen), it can be combined in solid metals, and it can mix with other substances to form gases.

Getting at it does not involve digging deep into the earth, disrupting rivers or cutting down the lungs of the planet. The waste product from using it is pure water, not climate-damaging carbon dioxide. And if the logistical puzzles of making and storing hydrogen could be solved, theoretically every household on the planet could make its own energy in the backyard from water. This is a vision of the democratization of energy that the American futurist Jeremy Rifkin, among others, is putting about.

THIS QUEST FOR social equality or for preserving the planet's life forms is not what drives Iceland's hydrogen guru Bragi Arnason. The white-haired, ruddy-cheeked head of chemistry at the University of Reykjavik is not convinced of the scientific basis of climate change and stoutly refuses to become involved in that issue. He is simply a pragmatist, like the determined Icelanders of legend. For years, he has been reading the scientific reports on how much oil and gas is left on the planet. He's in no doubt that oil and gas production across the planet will have peaked in the first few years of this century, and that energy shortages will come shortly after. He's just trying to make sure that Iceland will still have a source of energy when the oil goes dry.

Arnason eyes me carefully as he tells me this in his

cinder-block office at the university. The idea that the world is about to undergo an irreversible decline in oil production is heretical, as he knows all too well. He lectures about this all over the world, and constantly runs into people who believe that the oil will never run out, who believe that if they keep drilling, they will always find more oil. Talking about the end of oil is taboo. He shrugs. His white hair is so fine I can see the ruddy skin of his scalp underneath. He smells slightly of the cigarettes he loves to smoke, and his shirts are a trifle tight around the belly. He's not an activist. He doesn't need everyone to believe him. He just knows that it's a strategic advantage to understand that oil will run out. Soon.

He reminds me of Darwin kicking up against one of the most dearly held beliefs of his society. Even in a nation of free thinkers, Bragi Arnason stands out. Only, this time what's at stake is not the fixity of species or of today's ecosystems, but the fixity of ecosystems long gone. It's about an almost magical belief in the ability of the earth to provide whatever humans want, whether from what it produces today or from the remains of what it produced long ago. Today's petroleum products are simply plants and animals that died out as much as hundreds of millions of years ago, caught by geological accident on a seabed and then trapped by sedimentary rock on top. Eventually, these dead creatures were cooked by high pressures and temperatures into oil and, at even hotter temperatures further down, into gas. But it's a relatively rare phenomenon, occurring only when unusual circumstances coexist. And when the remains of these ancient ecosystems are gone, they are gone. There won't be any more petroleum for millions more years, until today's creatures have undergone a similar metamorphosis, if the conditions permit.

I THINK BACK to the Azraq Oasis in the deserts of Jordan. Vibrant for 250,000 years and killed in thirteen. How different is this from using up the earth's cache of oil, formed over hundreds of millions of years, in just a century?

Arnason is not the only one to break the taboo and talk about this. The first credible scientist to raise the warning flag was M. King Hubbert, a brilliant and bellicose scientist who predicted at a meeting of the American Petroleum Institute in San Antonio, California, in 1956 that US oil production would peak in the early 1970s. Up until five minutes before his lecture, employees of the head office of Shell Oil tried to persuade him to keep quiet. He refused and was roundly vilified for years.

The excoriation abated in 1970 when United States oil production peaked and then began to fall, just as Hubbert had predicted. Today, more than half the oil used in the United States comes from other countries. Beginning in about 1995, other scientists have tried to predict when world oil production will peak, using Hubbert's methods. They have estimated that it will peak sometime between 2003 to 2008 and then drop off permanently. This is still hugely controversial. At the same time, demand for oil will rise as both the world's population and its appetite for energy increase. In *Hubbert's Peak: The Impending Oil Shortage*, published in 2001, Princeton University geologist Kenneth Deffeyes wrote: 'It looks as if an unprecedented crisis is just over the horizon. There will be chaos in the oil industry, in governments, and in regional economies. Even if governments and industries were to recognize the problems, it is too late to reverse the trend.'

Arnason looked at the numbers in the 1970s and came to the same conclusions. That's when he started researching other forms of energy. Hydrogen emerged as the obvious choice because it was potentially the cheapest, cleanest and most plentiful. All you need to produce it is water and electricity, which in Iceland comes from the steam of the underworld. Liquid hydrogen is already in use as the main fuel of the space programme. And airlines have long since realized that it is the ideal fuel for jet engines because using it gives off no heat; also, liquid hydrogen is far lighter than petrol fuel and so takes less energy to transport within the aeroplane itself.

The breakthrough Arnason was looking for came in the early 1990s with the invention by Geoffrey Ballard in Canada of the hydrogen fuel cell, which transforms hydrogen into electricity by separating the ions from the electrons using a polymer membrane. The cell works something like a battery, except that the chemical reaction doesn't use up the cell. Instead, it uses up hydrogen piped into it, converting the chemical energy contained within the hydrogen atom itself straight into electricity. No burning. All it needs is a continuous supply of hydrogen, the most plentiful element in the universe and the one with the simplest atomic structure.

Once Arnason heard of the fuel cell, his labs in Reykjavik terminated all their tests on other mechanisms for using hydrogen and regarded the fuel cell as the engine of the future. In fact, Arnason has no hesitation in declaring that hydrogen will be the main energy source of humankind in the second half of this century. Iceland, in concert with a raft of other European cities, is already conducting a pilot project to run city buses on hydrogen gas. And Iceland has plans over the coming ten to twenty years to convert cars and ships to hydrogen too. In a couple of decades, Iceland will need not a single drop of oil.

I have sat here in Arnason's office, listened to his patient explanations and seen the shiny prototype run a small fan in his lab. I am deeply sceptical, mostly because it seems like such a smart idea and I want it to work. This is still a gestating technology and may be stillborn, I say. Can you really take it from this lab to the car in my driveway? He shrugs again. Of course it will work, he tells me. It's only a question of time.

I leave his office wondering why the switch to hydrogen is happening here, of all places on earth. What is it that allows this society to deconstruct this part of the fable of the earth's inexhaustible bounty? I wander around Reykjavik, trying to read the people. The posh downtown smells of sulphur and fish, a reminder of the fire down below and the water all around. Icelanders make use of both by firmly believing in the medicinal qualities of poaching themselves in salty hot springs. Some of the most prominent signs on Reykjavik's streets indicate the location of outdoor springs. They are crowded with bathers of all ages long into the evening.

I bathe along with them, sussing them out. The answer must lie in the legends of this close-knit, highly literate, highly educated people. They must be raised from childhood to innovate, to strive, to question. I have to explore this. I have to go to the north of Iceland, where the myths are even more alive than here in the capital.

SIGRIDUR SIGURDARDOTTIR picks me up at the airport in Akureyri. She is one of the many ethnologists of Iceland, an expert in the study of her people and their ways. She runs a clutch of museums along the north coast of Iceland, a stone's throw from the Arctic Circle. We are going to see the remains of a house lived in more than a thousand years ago, just discovered in an archaeological dig. The house might have belonged to one of the legendary saga clan members, Snorri Thorfinnson, reputed to be the first European baby born in North America.

Of course, it's not clear that he ever existed. No one really knows whether these magnificent sagas written in the Middle Ages are history or fiction, or a mixture of the two. But to Icelanders, Sigurdardottir tells me, they carry deep meaning, if not historical truth. It is something like the hold the Arthurian legends have on the British.

As we drive along the rugged coastal road, past hillocks where the elves are said to live, she tells me Snorri's story in language so rich, so detailed, so fluent, that I feel linked at that moment to all the other rapt listeners who have heard this story told over the centuries.

It starts in the tenth century with a clever, rich young farmer called Thorfinnur Karlsefni, whose name means that he was fated for big things. At that time, only a few people knew how to sail between Iceland and other parts of the world, but Thorfinnur, a sailor-merchant whose parents were wealthy farmers on Iceland's northern coast, was one of them. One day, he set sail for Greenland, where he met a beautiful and wealthy young widow, Gudridur Thorbjarnardottir, and fell in love with her. She had been married to Thorsteinn, the oldest son of the notorious Erik the Red. Erik's younger son, Leifur Eriksson, had just returned to Greenland from finding the New World, so Gudridur and Thorsteinn set sail too, eager to find it. But they got lost and failed. Eventually, they made their way back to Greenland and Thorsteinn died shortly after, leaving his share of Erik's fortune to Gudridur.

When the adventurer Thorfinnur asked Gudridur to marry him, she leapt at the chance. Very soon, they, too, set sail for the New World. This time, the jubilant Gudridur found it, including the houses Leifur had built. They stayed for as long as three years, discovering much more of the new land than Leifur had before them. Their first child, Snorri Thorfinnson, was born there. Eventually, they sailed back to Europe, their ship heavy with wondrous goods from the New World. They sold these in Norway and made another fortune, then sailed home to Iceland, hoping to be warmly welcomed by Thorfinnur's parents.

It was not to be. Thorfinnur's haughty mother did not take kindly to Gudridur's low birth and made life miserable for her. So Gudridur and Thorfinnur bought Glaumbaer, their own farm. This is where Snorri lived and farmed. When Gudridur became a widow for the second time, her wanderlust called out again and she set off on foot from Glaumbaer to visit the Pope. She promised that if she got back to Glaumbaer alive, she would build a church there for her God. But by the time she got back home, Snorri had already erected the church in honour of his intrepid mother.

Sigurdardottir tells me this story with relish, her eyes flashing, her eyebrows arching with every fresh twist of the tale. I think of Sarah Kuptana, the eldest of the elders of Sachs Harbour, halfway across the top of the planet from here, telling the story of how her father resurrected Phillip Haogak. But Kuptana bore witness to the events she told me about, and her tale had no shade of the fictional epic. Sigurdardottir's story is a thousand years old, told and retold over many generations. Sigurdardottir is deconstructing it even as she tells it, puckishly inviting me to choose whether to believe it or not. She knows that its truth lies not in the details but in the force of the collective imagination that has the power to believe it.

By the time Sigurdardottir has finished telling me the story, we are at Glaumbaer in a farmer's field, standing on top of the archaeological dig that has uncovered what might be Snorri's house. Not far away, the first Christian church of the area, with graves dating back to 1030, looks west to the sea and the New World.

Perfectly preserved even now, Snorri's house is built of turf, the root system of marsh plants that Icelanders cut from the ground with a spade or scythe in thick, decorative pieces. Turf was the prime building material of the Icelanders for 1,100 years and a perfect insulation against all extremes of weather. Icelanders figured out how to use turf after all the big trees were cut down en masse. The trees were never replanted. Adept at employing whatever lay at hand, Icelanders also used turf in place of saddle blankets on their pony-sized Icelandic horses.

The house that might have been Snorri's is massive, and obviously belonged to a grand family: 29 metres long, walls a metre and a half thick. Nearby was a mine of bog iron that the farm's inhabitants could smelt on the hearth. Sigurdardottir is the curator of the sprawling turf farmhouse up the hill a few hundred metres away, which replaced this one sometime before 1104. She takes me through it to show me the habits of Icelanders who lived in houses of turf. It is dark and dry and alive, like being inside the earth itself.

Most of the living area was given over to making and preserving food. Icelanders wasted nothing. Under the counter of one pantry are two massive barrels containing the sour whey left over from making *skyr*, or curds, from milk. Icelanders would throw scraps and anything inedible into the whey, where it could keep, pickled, for years before eventually being eaten, Sigurdardottir tells me. Even meat bones went in the barrel, the hard calcium eventually dissolving in the whey's acid, leaving a tasty morsel of sour marrow for a later meal. Today, most modern Icelandic kitchens have a small version of the whey barrel, and Icelanders have an enduring taste for the extremely sour and the wretchedly rotten.

THE BEST EXAMPLE of this Icelandic propensity for making something out of nothing obvious is the Blue Lagoon, a spa in the island's southwestern peninsula. Arnason has told me that I must travel there to see it. So I arrive at what has become by far the most popular tourist attraction in the country, complete with its own line of pricey skin-care products and a reputation for its ability to heal skin disorders.

The great joke is that this salty, healing spa is the waste pit for Swartsengi, the world's first geothermal electricity and space-heating plant. Swartsengi's stacks, yawning round windows and Star Wars architecture form a backdrop to the spa.

The Icelanders have long realized that this peninsula is the top layer of an underground radiator fired by heat from the earth's crust. Opaque steam rises constantly over the volcanic landscape, hissing up to the surface through fissures and faults. For generations, Icelanders have baked bread here on the peninsula by burying it half a metre down, covering it with gravel and leaving it for twenty-four hours. Thorsteinn Jonsson, who works for Swartsengi, has taken me on a tour of the peninsula. As we wander through the eerie, barren landscape, he points out a metal oven still in use, wedged up against an active fissure.

The inner workings of the earth are this evident here because Iceland is one of the newest bits of the surface of the planet, created only 20 million years ago and still a hotspot of volcanic activity. In effect, it is still being born. I pick up a reddish piece of volcanic rock, and Jonsson can tell right away that it's from the eruption of 1226.

As well, Iceland sits on top of two tectonic plates – the Eurasian and the North American – that are splitting apart at the rate of two centimetres a year. That means the island has earthquakes nearly every day. It's a phenomenon the Icelanders cheer over, Jonsson says, because the quakes open new fissures. And that allows the ocean to move deep inside the earth, transforming into steam from the heat and absorbing minerals from the surrounding layers of rock.

In the 1950s the Icelanders started to drill a kilometre or two down into the earth to tap the hot, salty water. It burst out at about 242°C. They became the first people in the world to figure out how to strip the energy from the water and convert it into electricity. Once they extract enough heat to get the water down to 100°C, they transfer some of that heat to fresh water and run it through insulated pipes into houses and other buildings in Iceland.

Today, about 87 per cent of Icelandic houses and industrial plants are heated by inexpensive geothermal energy from this plant and two others. And once the hot water heats up their homes, they run it underground to keep pavements and driveways free from snow and ice.

The Blue Lagoon was formed when the sea water, stripped of some of its thermal energy but otherwise identical to the way it came from the earth, was piped out the back of the plant as waste. Scientists thought the porous volcanic land would just absorb it. Instead, the minerals it contained sealed the pores and created an ever-expanding hot pool, potentially an ecological disaster.

The Icelanders, with their love of hot springs, didn't see it that way. They began flocking here to bathe in the healing fluids, tinged blue because the silica in the water absorbs the colour red. Eventually, the lagoon became such a draw – 300,000 visitors this year – that the people who run Swartsengi moved it slightly further away from the plant and built a spa around what is now a 6,000-square-metre pool.

Today, the healing powers of the lagoon water are so well documented medically that both the Danish and Icelandic medical systems send people here to heal themselves in the salty fluids.

I CAN SEE WHY. I have been soaking for a couple of hours in this primeval pool, rocked by the shifting of the earth under my feet, cradled by both water and fire. My hair grows stiff with the salt. My old skin sloughs off, the new blooms in its place. I am shriven. The steam rises so thick that it is impossible to see who else is in this place. I am blinded, able to move only by feel. Shapes become outlines. Male and female are indistinguishable ghosts. The smell of the salt fills my nostrils, seeps into my pores.

Suddenly, the steam clears, and I see a woman ducking into this cauldron to scoop up some of the healing mud and spread it on the face of her young daughter like a potion.

Then buses arrive at the Blue Lagoon directly from Keflavik International Airport. Travellers flock into the lagoon, stiff from hours of transatlantic flights. I can see the muscles of their backs relax as they penetrate the water. They groan with relief, with the joy of being surrounded. It strikes me then that we humans have not left our past behind. It is encoded in our DNA. Each of us carries with us the origins of humankind, of the first forms of life on this planet. The primal materials of earth are in us, and each of us is connected by these strands of genetic code to everything else that's here.

EIGHT

Ocean's End

There is life before you see the Galapagos Archipelago and life after, and they are not the same. Charles Darwin discovered the transformative properties of these islands when he journeyed here in the nineteenth century. Many voyagers have discovered them in the generations since, and I now count myself among them.

The scorched lava that forms this archipelago on the equator 1,000 kilometres out into the Pacific Ocean resembles a lost planet. It is a place of dreams and ideas, the end of the journey rather than the beginning of one, the spark to the writing of new legends.

The Galapagos have tempted me for years. They tempt me now to summon Darwin again, to ask him to teach us – again – the lesson of how life started and therefore of how it ends. Now I want him to tell us not the origin of species, but their fate.

Darwin's mark is evident here. The islands are redolent of him in the titles of places, in the reverence with which his name is still uttered. It is in all senses his archive, the archive of evolution, of the cracking of the great mystery of how life came to be.

For conservationists, coming here is like making a pilgrimage to the promised land. This is the one place on earth where financial resources, political will and international concern coalesce to ensure that damage done by humans will be overturned as much as possible, that the eccentric life forms that evolved here before humans discovered the islands will not be killed off.

That alone puts the Galapagos Archipelago in a league of its own. It is one of the most important and complete island ecosystems left in the world, a remnant of what was once a vast global collection of vibrant, whole and unique assemblies of life on islands. Most of the others are gone now. Local extinction spasms on islands happened in lock step over the centuries with the arrival of humans: New Zealand, Hawaii, Madagascar – these are just a few of the ravaged islands.

The Galapagos was headed in the same direction. Already, some key sub-species of giant land tortoise have gone extinct at the hand of man; others thrive only because scientists rear them in nature reserves. The Galapagos system as a whole survived because it is so isolated, so barren, so near the edge of what's possible for life. And because it is the stuff of legend. What would it mean if human society sat back while Darwin's laboratory – the one that explained us to ourselves – were sacked?

I HAVE COME HERE to live for a week on a boat with a team of scientists associated with Washington-based Conservation International. These professional dreamers, as I think of them, are working to preserve the diversity of life on earth. The government of Ecuador, which controls the Galapagos, has just changed – again – and the CI staff is determined to ensure that its plans to preserve this region's disappearing life forms will survive the new leaders. Last night, they organized a dinner party in the Ecuadorean capital, Quito, with Edgar Isch, Ecuador's new minister of the environment, and an international cast of conservation luminaries and got his enthusiastic – and very public – seal of approval on their plans.

Today, as we set off on-board the *Daphne* with the Ecuadorean crew stiff and formal in their starched white suits, CI has invited the cream of Ecuador's conservation bureaucracy to travel with us: Marco Altamirano, who has been director of the Galapagos National Park for just four days, and Fernando

Espinoza, executive director of the Charles Darwin Research Foundation. A large part of the reason they're here with us is Sylvia Earle, probably the most famous undersea explorer alive today. She's in charge of CI's new marine programme and is explorer-in-residence for the august National Geographic Society. She is a passionate and meticulous chronicler of the damage modern society is doing to life in the oceans. She is, in a sense, the marine equivalent of Darwin, exploring and describing an unknown natural world and its significance to a ferociously disbelieving public.

Earle is my bunkmate on the tiny boat. This is my first extended time on any boat other than the ocean liner I crossed the Atlantic in when I was thirteen. But Earle, who has explored the seas for three decades, has lived a great deal of her life either on the water or in it. She instantly puts her clothes away in what is obviously a shipboard system – I notice they are all black – and kicks off her shoes. It's better for balance, she explains. I kick mine off too. Riding the vagaries of the ocean – even in a yacht – works muscles in the feet and toes I never knew were there. She starts plugging in all sorts of electronic contraptions to charge their batteries. I am baffled and awed by them. Finally, I realize that one is a set of electric curlers, and all of a sudden I relax.

YOU HAVE TO BE PATIENT to enjoy the oddities of the Galapagos. You have to search vigilantly, quietly, and trust in the ancient patterns of nature. It is, as Darwin wrote in his journals, a reptile paradise, which means that tortoises and iguanas and other cold-blooded creatures fill the space that warm-blooded mammals occupy in most of the planet's ecosystems. These hardy reptiles were able to travel the great distances from the continent millions of years ago when other classes – mammals and amphibians, for example – faltered. Once they got here, they were able to survive the virtually barren conditions. Being here feels eerily like a return to the grand age of reptiles, when dinosaurs ruled the earth and mammals lurked at the edges of life.

We are searching for a specific reptile on the coast of Isabela, the biggest island: the mythical marine iguana. Darwin disliked them, calling them 'large (2–3 ft), most disgusting, clumsy lizards'. He notes, peevishly, that he has heard them justly called 'imps of darkness', a slur on the islands themselves, which he likens to the bowels of hell. Marine iguanas are endemic to the Galapagos, which means that, like so many other life forms on these islands, they evolved exclusively here and nowhere else on earth. They are the only lizards in the world whose livelihood depends on their going into the ocean for food. All other types of lizards live only on land.

Earle points. Tracks in the sand: footprints on either side of a shallow trench, the sign of the four-footed iguanas trailing their stocky tails behind them. And then, there on cindery boulders of lava rock on the edge of the Pacific Ocean, the imps of darkness. They look almost as craggy as the lava, heads studded with sharp bumps. Stiff spines – like the tines of many forks – curve down the length of their backs and tails. They look like mini-dinosaurs.

One big fellow captures my attention. He's not all black. His scaled skin is also shaded with subtle oranges and dusky greys. He's splayed on the black lava, soaking up heat from the sun, trying to get his internal temperature to a level that will let him move again. He's probably just out of the sea, where he was diving for algae to eat. As I watch, he looks at me and shakes his head. Then there's a short snort – a spray of pure salt, the creative way his body has adapted to getting rid of the huge quantities of salt he consumes along with the algae that make up most of his diet. The sun is blistering, even here at the edge of the ocean. It is extraordinary to me that any creature can live here on bare lava with what Darwin called the equator's 'vertical sun' beating down. No wonder Darwin came to understand from these islands that life is a constant struggle for survival, that the fight to pass on genetic material depends on the selection of a mate. Because so few places on the planet are this hard to live in, few places reveal the struggle so nakedly.

*

ALL BUT 5 PER CENT of the Galapagos Islands is a national park. About ten thousand people live in the limited area that is not park. And, as befits the most famous nature reserve in the world, the rules for visiting it are strict. The first one is that you can't set foot anywhere on park property unless you are accompanied by a guide who is registered by the national park. It's like having your own private park warden to make sure you follow the rules. Ours is Xavier Romero, a tall and muscular master diver who has a Ph.D. in biology from Ecuador's University of Guayaquil and who did one of his three degrees in Scotland. He still talks about the bone-numbing Scottish cold. I chuckle to myself, realizing that to him the cold was as much a hell as the Ecuadorean heat was to Darwin.

Romero rhymes off the rules. You must not use a flash, which would startle the wildlife, or put your camera too close or touch any animals or interfere with them in any way, or take anything, dead or alive, fragmentary or whole, out of the park. You must walk only on the trails marked for visitors' use. You must not pee in the wild, even in dire emergencies. You must act as much as possible as if you are not there at all. The rules are for everyone. And the visitors, even more than the local Ecuadoreans, are apt to patrol for infractions from other visitors. This is one place on the planet where humans play second fiddle to nature.

We pile into the dinghies – known as pangas here –

off the shore of Isabela to see what we can find where the sea meets the land. The air is thick with the taste of salt, the smell of rotting sea life, death and resurrection. It's a feast of rare wildlife sightings. Nearby are forty to forty-five blue-footed boobies, an international symbol of the Galapagos, named after the Spanish word for clown because they are so ungainly. A little further away, sitting on a piece of barren lava rock, is one of the peculiar black Galapagos cormorants whose wings have evolved into stumps no longer capable of flying. It's a stark reminder that nature eventually does away with whatever is not needed. These flightless cormorants are one of the wonders of the natural world, found only in the Galapagos.

The flightless cormorant is endangered now because the rats and pigs that humans have introduced relatively recently in evolutionary terms prey upon its eggs. It's what's known in scientific circles as the introduction of alien invasive species, a phenomenon that is spreading like a plague across each continent of the planet, killing off natural populations of plants and animals. Evolution, sedate and cautious when it came to doing away with the cormorant's wings, doesn't move much faster to develop protection for its eggs against new dangers – maybe camouflage or rat-proof shells. At the last official count, there were just eight hundred pairs of flightless cormorants left on the islands, which is to say, in the world.

We watch this one, the panga putt-putting over the noise of the surf. The cormorant stretches its sinuous neck, flaps its incongruously small wings as if to keep in practice and then tucks its beak under one of them. Finally, it dives into the sea, so graceful it looks as if it is in flight under the water.

Over on another outcrop sits one of the rare Galapagos penguins, the only penguin in the world to live in the tropics. It's tiny. More like a duck than a penguin. It even seems to swim on its belly, head in the air, like a duck. It's endangered too, with a total population of just 1,500.

And finally, a tall grove of red mangrove, a fantastical web of barnacle-encrusted roots twisting from the land into the sea. Mangroves make up one of the richest ecosystems in the world, and one of the most casually and efficiently destroyed. They're important because they are able to grow their roots in salty water, a rare skill in the plant world. The roots then capture silt and vegetation. Eventually, the coast itself expands into the ocean. That makes mangroves the cradle not only of new land, but of new life.

We sail quietly into a cove no bigger than our panga. The mangroves are giving off the light smell of citrus in the brilliant heat. And here are eight massive sea lions, draped over the mangrove roots, snoozing. It takes me a while to spot them all. Some are hidden deep in caves made from the stiff roots. But we are close enough that we could touch them if the rules allowed and if we had the courage. I notice that underneath their webbed flippers the skin is thick like old leather from their cumbersome journeys from water onto land. Their eyelashes are unconscionably long. One of them opens his eyes languorously and peers at us. We're not interesting enough to interrupt his nap. He exhales in ecstatic contentment and closes his eyes again.

This is one of the tallest mangroves in Galapagos. I can see its roots reaching for the sea like tentacles, propping and propagating at the same time. A thousand kilometres away along the South American coastline, mangroves like this are being torn down to make sea ponds for shrimp farming for the export market. The problem is, shrimp in the larva stage need mangroves to live. Once they're torn down, the shrimp won't grow any more. The market is gone and so is the mangrove.

A sea lion swims over to find out what's going on. Underwater, its bulky body glides effortlessly. It does a joyous dance in the sea and then raises its head out of the water to look at us. We can hear it take a breath. A little further off, we hear a baby sea lion calling. It sounds like a goat.

I look deeper into the water. Green sea turtles, taking a leisurely swim. A spotted electric eagle ray. More penguins. Schools of fish waving yellow tails. And that's just what I can see near the surface. This is a place of contemplation. My companions and I fall silent, watching, smelling, listening, feeling ourselves uncoil. Then one of the Ecuadorean scientists tells a story.

In the 1980s, some businesspeople wanted to build a five-star casino near here and put all these animals in a zoo. The proposal was under serious consideration for a while but was finally rejected. It's hard to imagine this timeless womb of creation broken by the short-term frenzy of a tiny proportion of the world's humans gambling, drinking, seeking ever more money, more thrills, oblivious to the future, to humanity's shelf life.

EARLE HAS TAKEN the top bunk in our cabin, leaving the wide low one for me. It's a gracious act because it means that every night she must shinny up a ladder to a far smaller sleeping space than the choicer one down below. It takes me a few days to realize part of the reason she's done it. Before she does anything else in the morning, and preferably before dawn, she greets the sea through the porthole visible only from the top bunk. It's a gentle ritual, but an important one. She is reading the strategy of the sea. I am reminded of Sarah Kuptana watching the frozen Arctic Ocean, trained in its subtleties. Where others see nothingness when they look at the ocean, Earle sees a complex ecosystem that is the foundation of life on earth. She watches it even when she seems to be doing something else, as if the salt of her blood is bound with the salt of the sea.

She started out like all other biologists: looking at

dead things and trying to figure out how they once lived. Finally, she balked. She wanted to study life, not death. But why the ocean? I ask her. Why not the land?

She is in our cabin, hanging from the ladder to her bunk, and swivels to face me. Because, she says, she's a biologist, and a biologist studies life, and most of the planet's life is in the water.

I'm obviously looking confused, so she climbs down and starts to explain. If you looked at the earth from space, she tells me patiently, you would think that it is a planet of water, not of land. Humans don't tend to think about this because we are terrestrial. We think the land is more important to us than the ocean. But we are wrong. The ocean is not only the original source of all life on earth, it also shapes the climate, the weather, the temperature, the chemistry of the planet and therefore the home of terrestrial species. Like us. In fact, 95 per cent of the earth's biosphere is in the ocean. Humans and all other life forms are utterly dependent on oceans. And how do we care for the planet's main life force? By treating it as the ultimate sewer and store, vacuuming up whatever we can lay our hands on, whether we need it for food or not. We are a shockingly stupid predator, killing not just far more than we need, but also the life-support system that provides us with so much food.

She delivers this passionate speech in the restrained voice she might use to give an elegant lecture at university. No wild gesticulations. No pounding of fists. Just an uncanny stream of logic and facts.

But I know the outrage is there. On board the *Daphne* I've been reading one of the books Earle has written, *Sea Change: A Message of the Oceans*, with its plea for humanity to recognize the damage and stop it. Earle has seen the destruction first-hand, one of the few scientists to have done so. She has seen whole ocean ecosystems laid to waste by fishing practices that assume the sea is infinite, that its riches are inexhaustible. But Earle has also been to the bottom of the abyss. She can see – even if others can't – that there are limits to what the ocean can provide. Already, 90 per cent of each of the biggest fish species in the world – tuna, cod, salmon – have vanished under heavy fishing in just the past five decades. The biggest threat to the ocean – and therefore our own future existence – is ignorance.

She's edging out of the cabin, heading for the deck and the store of magnificent coffee she's brought for the whole crew from Peet's, a specialty chain in California, aware of the shortcomings of South American boat supplies. I lob my perennial question at her. Are we a suicidal species?

Probably not, she says, sprinting up the ladder to the deck. We have the capacity to learn and to act in our best interests. It's just a question of more people figuring out what those interests are, biologically speaking.

*

I'VE BROUGHT DARWIN'S journals with me and an excellent water-proofed Libri Mundi map of the islands and the surrounding seas that I picked up from some thoughtful hawkers at the airport at Isla Baltra. I am compelled to read Darwin again as geographically close as possible to where he was when he was writing. As well, I need to sort out which of the archipelago's dozen or so islands he visited and what he found on each. These are probably the most analysed travel journals in the history of science, and it has become clear to me that there are Galapagos junkies in the science world who can say exactly which species Darwin sighted where, which craters he climbed in, which beaches he walked. You can't talk generically about the Galapagos after you've been here. You have to talk about which 20-metre stretch of which island.

Once I spread the map out in the centre of the common room, I realize we're retracing his steps for most of our voyage. Some of my companions – the scientists from CI, the donors from America – gather around the map in the sticky heat of approaching night to track his trail too, leaving their piles of books and computers and photographic equipment aside for the moment. We will be seeing some of what Darwin saw more than 150 years ago. And as we piece this together, I realize that Darwin's life and the lives of each of us on this boat are marvellously linked. We are here only because Darwin happened to get on board the *Beagle*, and the *Beagle* happened to stop here, and he happened to figure out his theories, publish them and convince the public that he was right. Of all my journeys over the past three years, I have never felt Darwin with me as strongly as I do now.

I start to read bits of Darwin's journal out loud, telling his tales of the islands in snatches. The Galapagenos among the crew, for whom this archipelago is simply home, watch us silently, inscrutably, as they go about the work of running the ship. Their pristine white suits have vanished. Today they are in shorts and T-shirts, barefooted like the rest of us.

Darwin disliked these islands. He'd been at sea nearly four years by the time he set foot on them and had explored many of the secrets of South America by then. He expected to be entranced by the geology and was hoping to witness some live volcanoes. He was cruelly disappointed in most of what he saw. The trees were stunted and seemingly lifeless, he wrote, after having spent his first hour on the islands. The plants smelled bad. And the black lava rocks were heated up like an oven by the rays of the vertical sun. In all, 'The country was compared to what we might imagine the cultivated parts of the Infernal regions to be,' he sniffed.

And that was Isla San Cristobal, then known as Chatham Island, one of the oldest and most geologically settled of the islands. By the time he made his way to Fernandina (which he knew as Narborough), the youngest and most volatile island, his opinion seemed to have slid. He said it presented 'a more rough & horrid aspect than any other; the Lavas are generally naked as when first poured forth'.

And Isabela, the large, seahorse-shaped main island of the archipelago, formed from the joinings of six volcanic eruptions from deep beneath the sea, Darwin thought simply sterile. 'I should think it would be difficult to find in the intertropical latitudes a piece of land 75 miles long, so entirely useless to man or the larger animals,' he wrote.

The days were sweltering. The relentless heat of the sand pierced the soles of thick boots. Water was in critically short supply. The flowers were ugly and insignificant, more suited to the Arctic than the tropics, he wrote. Nothing was lush. And the animals were so unused to humans that they unwisely considered them harmless. Darwin told of pushing a Galapagos hawk off a branch with the butt of his gun, of a shipmate who killed a bird with his hat, and of a boy on Floreana Island who killed one endemic Galapagos dove after another simply by sitting at a well with a stick in his hand.

All the elements of the Galapagos that Darwin found so unnerving were actually signs of the adaptations these creatures had made as they struggled to survive the wretched conditions of these equatorial islands. The trees, for example, are mostly one-sex, the ultimate survivors. That means they can propagate themselves without cross-fertilization from a tree of the opposite sex. And that, in turn, means that they do not need lavish flowers or gaudy colours or sweet scents to attract birds and bees to spread their seeds. The trees do not waste energy or adaptive powers developing frills not needed for the game of reproduction. This is a pared-down system of life, not the fabulously complex one of mainland South America. Rather than several species to fill each ecological niche, there is often just one.

That makes the function of each species critical to the workings of the whole. Small changes make big differences.

I think of the marine iguanas, those rugged salt-spitters. During an El Niño year, when the ocean water heats up slightly, the iguanas do badly. They still do everything they usually need to in order to survive: store heat from the sun in their bodies and use it to dive into the sea, gorge on algae and then surface to soak up the sun's rays again. But the slightly altered temperature of the sea means that a different type of algae grows in it. The iguanas can't digest this type of algae. So they starve en masse with full stomachs. One of the looming problems for the Galapagos is that global climate change is spawning more frequent El Niño years; the effects on such a simplified ecosystem as this could be disastrous.

OF ALL THE PIECES of this island system that Darwin misread, his misreading of the giant land tortoises stands out. By a great irony, they, along with the famous finches that bear his name, were Darwin's first clue that species might not be fixed, that a single species had arrived on the islands and evolved into separate species to suit the surroundings of each island as they spread out and became isolated. In his ornithological notes, written the year after he visited the Galapagos, Darwin wrote that the locals could tell which island one of the elephant-footed tortoises came from by its form, the shape of its scales and its size. It flabbergasted him. Why would the tortoises on different islands so close together be different? Why would they not all be the same? It was clear to him that they all filled the same ecological place in the islands. Could the differences be variations of the same species that had altered in isolation from each other in response to changes around them? That would mean creation was continuous, not the result of a single God-made burst. And that would be heresy.

'If there is the slightest foundation for these remarks, the Zoology of Archipelagoes will be well worth examining; for such facts would undermine the stability of species,' he wrote.

While he was here on the islands, though, Darwin saw the giant tortoises mainly as a source of food, not the inspiration for a scientific theory that would change the way humanity thought of itself. Like the doves around the well on Floreana, the slow-moving giants were easy prey and bountiful.

On Santiago (then called James Island), Darwin spoke of the giant tortoises that 'swarm' near springs and hinted that he rode one: 'The average size of the full-grown ones is nearly a yard long in its back shell: they are so strong as easily to carry me, & too heavy to lift from the ground. . . . During our residence of two days at the Hovels, we lived on the meat of the Tortoise fried in the transparent Oil which is procured from the fat. – The Breast-plate with the meat attached to it is roasted as the Gauchos do the "Carne con cuero". It is then very good. – Young Tortoises make capital soup – otherwise the meat is but, to my taste, indifferent food.'

On Floreana, then known as Charles Island, Darwin recorded that the giant tortoises were the staple of the human diet, easier to catch than anything else. 'The main article however of animal food is the Terrapin or Tortoise: such numbers yet remain that it is calculated two days hunting will find food for the other five in the week,' Darwin wrote.

And they were food for more than the handful of Galapagos residents and the odd travelling naturalist.

By the time Darwin arrived, the Galapagos, out in the middle of the Pacific Ocean, had long been a turtle-meat market for the crews of whaling ships. They, too, caught the massive tortoises, sometimes killing them and salting the meat, getting more than '200 pounds' of flesh from a single creature. Sometimes they stowed the primeval reptiles in their ships' holds and slaughtered them later, counting on the tortoises' legendary hardiness to keep them alive for months or years without food, water or light. Records from whaling ships show that whalers alone took 13,013 tortoises from the Galapagos Islands from 1831 to 1868.

Darwin wrote of one frigate ship that caught two hundred of the massive tortoises in a single day's hunting on Floreana. He noted that the tortoises were getting harder to find. 'Of course the numbers [on Floreana] have been much reduced . . .' he wrote in his journal. 'Mr. Lawson [the acting governor of the Galapagos] thinks there is yet left sufficient for 20 years.'

Mr Lawson was wrong, as I soon discover. One of the perks of being on this particular journey to the Galapagos is that nearly every important scientific article ever written about the islands is either in someone's backpack or burned into his or her memory. As I lounge in the ship's common room, reading and trying to figure out what happened to the tortoises of Floreana, Roderic Mast, one of the turtle specialists on this trip, has run down to his cabin to fetch me a book that he just happens to have in his suitcase. It's the definitive scientific text on the creatures, *The Galapagos Tortoises: Nomenclatural and Survival Status* by Peter C. H. Pritchard.

It picks up where Darwin left off in October 1835. The sub-species (or possibly species) of giant tortoise unique to Floreana was extinct within fifteen years of Darwin's visit, and possibly within five, Pritchard has found. Pritchard wrote: 'There is no evidence that this population survived beyond the 1840s.'

Other types of giant tortoise have vanished at the hand of nature and man too. Those on Fernandina (once known as Narborough) and Santa Fe (once known as Barrington) are gone. A single male from Pinta (Abingdon) still lives. Called Lonesome George, he was taken to the Charles Darwin Research Station on Isla Santa Cruz in 1972 and has been the subject of decades of heroic efforts to breed him with other sub-species of giant Galapagos tortoise. All have failed.

Several of the other sub-species have come close to the edge of extinction but were pulled back by scientists who decided to breed them in captivity. None of the remaining eleven is free from danger. The entire population of giant tortoises here has risen in recent years as a result of the breeding programmes to perhaps 16,000 in total, spread out over the eleven geographically separate populations and all the stations set up to rear them. I read in Pritchard's book, though, that before humans found the islands, in the days when the tortoises ruled these lands, there were perhaps 250,000 of them.

ONE OF THE MAIN purposes of this expedition is for some of the movers and shakers within CI to persuade the organization that it must begin to spend more of its money on marine conservation. Like most humans, the ones who run CI tend to think the land matters most, and that's where their focus has rested. If this trip manages to broaden CI's scope, that could well persuade other conservation groups and national governments to follow the lead. This has been one of the keys to CI's success.

That means CI has to choose a high-profile marine project that has an excellent chance of succeeding. It's a tricky prospect. Most of the open ocean isn't controlled by any government or international agency, and that means it's hellishly difficult to protect. Not only that, but nations with rich marine ecosystems – especially the ones whose citizens are dependent on fishing – are notoriously reluctant to co-operate with each other.

But a marine corridor linking the islands of Galapagos, Cocos, Coibo, Malpelo and Gorgona, involving the co-operation of Ecuador, Colombia, Panama and Costa Rica, seems to have a fighting chance. It fits the bill of being a celebrity site, and all four national governments are making the right noises about working together, after some heavy lobbying by CI. Best of all, there's a strong scientific case, painstakingly assembled by some of the biologists on this boat, that the waters joining these four groups of islands are rich breeding grounds for fish, turtles, sharks, whales and other creatures critical to life in the sea. It's all the result of the sea currents – six of them – that come together in a rare convergence, flowing both deep below the surface and near the top. With the currents come chlorophyll and plankton, the food of the ocean masses.

But in the world of conservation, scientific justification is only one part of the game. The other is visual evidence. And that means this group needs to scuba dive in some of the most remote parts of the sea with film and video, sometimes as much as four times a day.

Getting ready to dive is a male ritual. Sylvia Earle, the only woman to dive, neatly dodges it by getting into her wetsuit in our cabin and emerging flawlessly prepared.

I sit on the railing of the boat's back deck, swinging my bare feet, watching this primal struggle. Laughing. I grew up on the North American prairie and have never worn so much as a snorkel. The dozen or so men grunt and moan, shoehorning themselves into the tight black suits. Often, the suits are already wet from earlier dives and refuse to co-operate. The sun beats down. The men swear. They bellow for lost gloves, goggles, flippers, weight belts. It doesn't occur to them to preen or suck in their guts. The sweat rises, pungent. Ecuadorean deckhands run over to hose them down with cool salt water, load them up with oxygen tanks and regulators. All this they need, just to survive for less than a single hour under the sea.

Then the cameras are lowered, gently, tens of thousands of dollars of high-tech equipment reverently handed from deck to panga, until finally the panga sets off with a roar to the dive site where the men will roll backwards into the sea.

At night, another ritual: the showing of the images. The lights in the common room go out. We sit on the blue couches, feet on the wooden table that is anchored in the middle, skin puckered from the salt of the sea, faces reddened by the sun. Legs, arms, hips comfortably joined. It's like watching replays of favourite sports matches. Praise is mandatory: Look! We got the hammerhead sharks. We were this close to those manta rays! Can you believe the colour of those corals?

THE NEXT DAY, we are on Punta Espinosa on Isla Fernandina, the newest and rawest of the islands. There was a famous eruption here in 1842 that extinguished Fernandina's population of giant tortoises. It also melted the tar on ships in the cove and made the sea so hot that sea lions jumped out of it to cool off in the cindery air. We can still see the black lava set down in the patterns it must have formed more than 150 years ago, within minutes of cooling off, the remnant of a timeless cycle deep within the earth. It seems to me that I can read in these undulating layers the strategy of that fire from long ago.

Thousands of female marine iguanas are nesting where the shore meets the sea. They are vying for the best piece of beach real estate to harbour their eggs, oblivious to the humans gawking at them. This is a ritual too. The females use their legs and claws to dig holes in the sand, gracefully tossing the sand to either side in time with an ancient beat that goes on and on. All over the beach, iguanas are doing the same thing. I can see short sprays of sand flying up everywhere I turn. I am witnessing the continuation of one of the species that evolved right here on this shore. I'm close enough to one of them to see her pink tongue flit from her mouth.

IT HAS NOT ESCAPED my notice that Earle is one of the very few female biologists to have become a celebrity. Thinking of Oxford, that irrepressible boys' institution, I ask her how she made such a name for herself in academia. She bristles. You just have to take stock of who and what you are and get on with it, she says, walking the shores of Fernandina in her bare feet, elegant in black. Pressed a bit, she remembers that at

Duke University, nearly forty years ago, when she was doing her master's degree, she was denied financial support because she was a woman. Her academic advisers told her that she would just end up marrying, staying at home and having children. Eventually, she found a part-time job as a herbarium assistant that paid her two dollars an hour for twenty hours' work every week. It wasn't much, but it was enough, she says grimly.

She did marry and have children. But she never stayed at home. Instead, she has set diving records for decades, has established a company to create submersible machines and has become such a renowned marine biologist that she is known as 'Her Deepness' in most important circles, outstripping most of the male biologists of her generation. And she has written four children's books on the sea, an unusual undertaking for a scientist of her calibre. Why? I ask her, as the iguanas continue their rhythmic birthing dig. Simple, she answers. If we don't inspire young people with things that matter, the things that matter will be lost. She stops, craning her neck to see a point on the lava rocks around the bend. Flightless cormorants. Eleven of them in one place. Almost 1 per cent of the world's population. Her eyes shine with wonder.

SHORTLY BEFORE the sun sets, we explore the older, northern part of Isabela. The water is topaz, the rock sedimentary, heaved up from the sea bottom thousands of years ago. One of the bucklings has created a cave, and we enter it in our pangas, tiny against the backdrop of this massive piece of the planet's crust, corroded underneath by wind and water. I try to see how this piece was ripped away under the sea and wrenched up on its side. Forces unimaginable to mere humans have made this. I had come to think of the earth's crust as stable, protective, growing stronger with each passing millennium. Now I realize that it, too, is changeable.

Our voices echo in the dark, crystal clear. There is barely enough light to see into the water, but gradually our eyes adjust. It is the stuff of every watery nightmare I have ever had. Rays gliding soundlessly, mere shadows under the boat. Menacing forms without shape, reaching into eternity.

EARLE, ESPINOZA and Altamirano have left, carried away on a sea taxi back to denser civilization. The rest of us have spent the night travelling to the most distant, most unapproachable islands of the Galapagos, where no human can set foot: Wolf and Darwin. By 6.30 in the morning, when I climb up on deck, we have already dropped anchor and are settled in a cove off Wolf. For the first time, we see red-footed boobies. Juveniles. Dozens and dozens of them perched on every available surface on the top deck of the ship, staring at us. They have flown over from the sheer rock face of Wolf Island to check us out. More fly over. And more, unsteadily using their spread-out feet as windbreaks to help them land. We could touch them if we dared. One of the diving masters, Juan Carlos Moncayo, who we know just as Macarron, has come to join the fun, holding out his arms as resting places for the boobies.

We are in the remote Pacific Ocean now, where other boats rarely come. We are all uncoiling, even the Ecuadoreans. It must be the bare feet, the visible strata of the earth, the sea, the sun, the wind. The long hot evenings watching video images from the secret world under the ocean. We feel more fully alive.

There are thousands of other birds on the island. Its rock face seems to seethe in the distance as they move. The air is heavy with their sounds. Big male frigate birds inflate the massive red balloons in their chests, drumming them to attract females. Some even try to fly with the unwieldy sacs still inflated, an excess of male pride that is doomed to fail. The females appear nonchalant.

Dolphins play off the side of the ship, waiting for it to hoist anchor again. They pretend to race it, as if warming up until it is actually moving again. When I lean over the prow of the ship and look straight down, I can see dolphins gliding joyously through the water.

THIS IS A HEAVY DIVING portion of the trip. Most of the men go down three or four times a day, including night dives in the inky ocean. When they are not diving, they are preparing to dive. One of them has found an abandoned piece of longline, a hellishly wasteful method of fishing that sets hooks kilometre after kilometre, killing whatever fish, bird or mammal happens by. It is shocking. This is supposed to be a sanctuary. Yet here we find evidence of the most destructive fishing method on the seas.

Our guides, Romero and Macarron, say that shark finning goes on here too. I've never heard of this. They explain. Shark fins are prized in Asian markets for soup and as folk remedies. They fetch high prices. So fishermen will come to remote parts of the sea like this one and catch sometimes thousands of sharks at a time, slice off their fins and tails, then throw the bleeding, living creatures back into the sea. Unable to navigate, they fall to the bottom and drown. Romero and Macarron don't see nearly as many sharks in the waters off Wolf and Darwin these days, they say.

THE ECUADOREAN crew has utterly thrown off formality. They are in bare legs and thin T-shirts. Now that we have come to the most remote part of their land, they are involved with us, making jokes, laughing, breaking bread, watching videos.

And flirting. The captain, whose name I never know, takes me aside one day while the others are diving or asleep and starts laying on the compliments. He offers to show me his steering wheel, his cabin lined with shiny wood and stern compasses. The bed where he sleeps. He has noticed me laughing. I am a free spirit, he declares, and what am I doing after dinner? I laugh and leave quickly.

Not to be outdone, Macarron offers to teach me to dive. He hooks me up to his regulator, and I float off the back of the boat in the middle of the ocean near Darwin, trying to breathe underwater. The salty sea invades my mask, fills my nostrils. I panic, terrified that I will drown, kick up against the engulfing ocean and gasp for breath. There's too much I don't know about how this works. For this, I do not have enough faith – this takes more faith than swimming with piranhas.

He doesn't give up. A day later, when we are in a shallow bay off the Isla Bartolome, he urges me to try again. We've just come from seeing scores of sea lions up close in the burning sun, and my ears are still filled with the sound of one mother nursing her young, the greedy sucking audible above the waves. If that is possible, what is not?

So we fix the leaky mask, spitting in it to make the seal tighter. And it's shallower water on top of white sand, not the deep black currents of the open Pacific. Now it's my dressing ritual, as Macarron and several of the other crew members stretch a wetsuit around me, plunge my feet into booties and flippers, fasten a weighted belt around my middle, hook up oxygen tanks.

I may have thrown off some of the shackles of civilization in the Galapagos, but I still find this deeply uncomfortable. Especially when two of the crew give me the glad eye and the thumbs-up.

But the embarrassment shifts to delight. I stumble into the panga, laughing uproariously, proud of myself. A stream of last-minute instructions on how to stay alive and avoid brain damage. And then, Macarron says, simply: your life is in my hands. Trust me.

What surprises me most is all the sound. I had imagined that the sea would be silent. But it is alive with sounds: trickles and booms and flicks. A stream of bubbles every time I clear my mask. Macarron holds my hand, guiding me, giving me instructions in sign language, clapping gently when I spot a sea urchin or a purple starfish. A school of yellow-tailed fish swims towards us. All of a sudden, we are in the middle of it, and the fish part to make a path around us. Below, just beyond the reach of the sun's light is a manta ray, dark and covert.

I am not afraid. It never even occurs to me. I forget that I'm breathing through a regulator, with oxygen out of a tank. I forget about the bends and the rules and all the strictures of gravity, both social and physical. It is as if I have entered a new world where joy and serenity reign.

It's time to surface. I've been 12 metres below for forty-eight minutes. It felt like so many seconds. I can't get the grin off my face.

*

BY THE TIME we reach the main Galapagos Islands again, it's as if things have reversed. Before, coming from Quito to the islands, we felt as though we were coming to wilderness. Now, coming from the far more remote Darwin and Wolf, we're returning to civilization. When we get off the boat at Puerta Ayora, the economic capital of the islands on Isla Santa Cruz, we feel the bonds of common rules snapping back into place. It feels odd to be surrounded by shops and people and other boats. A mere week and we're slightly unstuck.

This is the main part of the Charles Darwin Research Station and the heart of the tortoise-rearing operation. The big success story is the Isla Espanola tortoises. In 1977 this unique population was reduced to fifteen individuals, some of them in the San Diego Zoo. Now, there are 1,200, and they are capable of reproducing in the wild. They still get a lot of help. And need it. If the breeding programme were to vanish, so would the tortoises. The young stay at the tortoise nursery until they reach five years old, kept in cement boxes lined with lava rocks and covered with mesh to keep predators out. The cost for each tortoise raised to that age is $5,000 (US). This station has about six hundred, and there are several more stations.

I think about the biggest fish of the sea, their populations cut by 90 per cent over the past five decades alone. They were fished out, just as the tortoises were hunted out. What breeding recovery programmes does fate hold for the fish? Will it be enough to stop killing them? Will it take something with more finesse? Or will we simply allow the key species of the biggest ecosystem on earth to vanish?

The tortoise-breeding procedure is not complex. Research station staff collect eggs from the wild and from captive breeders, then incubate them in rectangular, two-handled plastic blue basins with a layer of sterile vermiculite on the bottom. A hair dryer does the trick for regulating the incubator. I pick up an egg that has been found to be sterile. It is rubbery and greyish white. I am amazed at how heavy it is.

The toughest tortoise to breed is Lonesome George, the only specimen left from Isla Pinta. He has no mate, but that doesn't stop the scientists. For more than thirty years, Lonesome George has been kept here, his every ejaculation encouraged and catalogued. Roderic Mast, the tortoise specialist from CI who spent part of his youth as a guide in the Galapagos, remembers the time he kicked off work all those years ago and went to a bar down the road with a friend. They ran into a pair of beautiful Italian women, students who were working at the research station for the summer. As the drinking progressed, and the talk carried on, the Italians revealed the nature of their job: manual stimulation of Lonesome George. In the name of science. It was a conversation-stopper, Mast recalls.

He's still here, is Lonesome George, over there in his own enclosure under the prickly pear. One research staff member confesses that he was getting fat and they had to put him on a diet. The search continues to find him another giant tortoise to mate with. They're trying some females from the population on Isla Espanola, but have had no luck so far. Lonesome George is only 70 to 80 years old, still in the bloom of youth compared with the 175 to 200 years that scientists estimate he could live. But, the staff member confides, things are not hopeful these days. He mounts but he no longer ejaculates. They've been keeping track. I have to wonder what Darwin, with his theory of natural selection, would think about all this.

We're on our own for a couple of hours to wander in the blistering heat of Puerto Ayora. Our park-appointed minders have gone away for the first time in a week. It's a little like being let out of a cage. We take a panga across the harbour to a beautiful café on the water. There are lights and music and soft breezes. We order wine and listen to the unfamiliar sounds of people. The fine tangerine silk of my shirt sticks to my skin.

AND THEN IT'S OVER. The last supper, the last on-board sleep, the end of the journey. We pack our bags, put our shoes back on. We are re-coiling. The alchemy is wearing off. The Ecuadoreans are all business again, getting ready for the next lot of paying passengers, and those crisp white suits and peaked caps have miraculously reappeared.

One of the scientists asks me where I'm bound for next, and I burst into tears. Hot, intractable ones. The ones that don't stop even though you desperately want them to. This is the end of the quest, and I am supposed to have found some answers. All I have found is that the land is in terrible trouble, but not as much trouble as the sea. And for the first time, I understand how the two fit together and why they must.

I have found Darwin, seen what he saw, consulted his oracle. Sought the lost islands and the promised land. But I have no big answers. No tidy new legends. Darwin's ghost cannot foretell the fate of the species because that fate is in our hands now. It is bound in patterns of human behaviour.

In the end all I have found is that the biggest part of searching is hope. And that if we give up hope, the quest has ended.

EPILOGUE

Writing New Legends

Quests never really end. A few months after leaving Galapagos, I journey deep into one of the last tracts of virgin forest left in the world: the boreal forest. Mantling the top of the planet below the tundra, the boreal forest has none of the grand sweep of the rainforest or its awe-inspiring structure. Many of the trees are stunted and spindly. Nature seems sparse, almost stingy here, without the luxurious abundance, the fervid production, the fantastical variety of forests at the equator.

But for me, the northern forest offers a subtle and profound beauty. You have to look closely to find the wild raspberries and blueberries or the trace of the woodland caribou and the moose. The glossy leaves of the kinnikinnick, or bearberry, a shrub that lives low to the ground in the shallow soil, are no bigger than my smallest fingernail. My family has had a wilderness cabin on a lake here for more than two decades, built by my father, George Mitchell, who was one of North America's pioneering biologists. And it is remote. In my family, we joke that to get there you drive to the top of Saskatchewan, keep going, and then take a boat.

The great task when we arrive is to see if the black bears have broken into the cabin over the winter. One year they managed to tear the screen door off its hinges and demolish the screen. Another year my brother ran into one at a stream across the lake from the cabin. He stiffened in shock and took off in one direction while the bear, equally shocked, took off in the other.

Normally they give us a wide berth. But the worry is not only that they will have found the cabin, but that they will have become possessive about it. We've heard already that it's been a rough year for the bears. They have been spotted further south at rubbish bins foraging for food. Bears don't like to be hungry.

So when we get to the sandy landing with our motorboat, we make a lot of noise, counting on our loud voices to warn the bears – if there are any – that we are about to arrive. Bears don't like to be surprised.

But all is well. The cabin is untouched. Off in the distance, the haunting song of the loons. Their voices carry across the deep, cold lake, weaving through the trees of the forest. It's an old song, here in this ancient forest, and it makes us shiver with its primal power. Everything is just as it has been for thousands of years.

Except that it really isn't the same. Like the Arctic, like Madagascar, like Jordan and Galapagos, this part of the world is changing on a grand scale year by year. After we lug the food, clothes and propane tanks up the hill from the landing to the cabin, install the new water pump at the lake's edge and settle down in the dusk for a glass of my father's home-made cherry wine, he starts talking about how it used to be.

When he first started coming up here more than twenty years ago, the roads were winding and rough and often impassable. He remembers one year when a pond formed across the road and you had to hook your car to a big-wheeled tractor to get across. Now the road has been straightened out and it's paved almost all the way to the cabin turnoff.

All the better for the logging trucks heading south from the clear-cutting in the boreal forest. For kilometres along the new road, the forest has been razed and turned into timber. Every single tree for hectare after hectare. Except for a fringe on either side of the road so drivers won't see the devastation deeper in. This is the nibbling away at the bottom edge of the boreal in Canada, and it is happening in Russia and the other northern countries that are home to this ragged forest. Oil and gas exploration is going on too, and mining exploration. That means lots of other new roads are eating into the forest. The final frontier, they call it. The last part of North America that waits to be broken.

My father notices a few subtle changes at the cabin already. There aren't as many squirrels as there used to be. And probably fewer eagles. It's hard to say whether that's a result of the clear-cutting, but it's something to watch.

There's still an enormous amount of boreal forest left. Large swathes of white-skinned birch and poplar and pine are still considered intact. It is not too late to save the forest, but logging, oil and gas, and mining industries have staked claim to a large percentage of the boreal. It has been done without taking into account the fact that this is the largest untouched forest in the world and that it is necessary to the healthy functioning of the planet's chemical and water systems.

The boreal may face worse than that, though. Because of its location in the middle of a vast northern continent, the boreal will be severely affected by global climate change. In another twenty years, what will happen to our family refuge in northern Canada? Like the natives of Madagascar, will we cling to the myth that if we keep walking we will find another tree?

What is it about humans that lets us see these trees but not the forest, not the life it helps to support?

I remember stumbling over a fragile and dusty volume the colour of old claret as I was poring over books at Oxford's Radcliffe Science Library, musing about Darwin. I had ordered it from the stacks on a whim, not expecting much.

It was a copy of a speech called 'Why I Am a Darwinist', delivered by Sir Arthur Keith, president of the British Association, at its annual meeting in Leeds on 31 August 1927. This was shortly after the Scopes trial in Tennessee had captured the world's imagination over Darwin's theory of the descent of man and apes from a common ancestor. The public was in confusion over what it thought of as 'the monkey theory'. Sir Arthur was clearly playing to a big audience.

He wrote: 'Until [Darwin] came men believed that the miracle of creation took place long ago, and in a distant place; he produced convincing evidence which shows that miracles have not ceased; they are taking place here and now; creation has been, is now, and ever will be a condition of human existence. Evolution is at work this day among us, and among all the people of the world.'

He ended with a prophecy, almost a foretelling, of the human reasons for the current ecological crisis. But it is also the blueprint of how to write a new legend for our times: 'We shall rise to higher things, and yet the one postulate which may be asserted with some degree of confidence is that man can never become a merely rational machine – one obtaining the zest for life by the exercise of higher mental faculties used in a purely unselfish way. The more we come to know the "makeup" of the nervous system of men and women, the more we realize that the deepest and most lasting pleasures of our lives are old instincts which are deeply implanted and of very ancient origin in the animal world.'

In a sense, Sir Arthur is saying that no matter how capable humans are of understanding science, science will never have the emotional force of legend. Humans respond reliably not to information, but to meaning. Not to knowledge, but to understanding. The imperative is this: tell me if you must, but also show me; this is the ancient way of animals. It is our evolutionary birthright.

We have proven ourselves capable of wholesale shifts in social values over time, in rewriting legends. Sitting up here in my family's cabin in the boreal forest, musing over my voyages, I'm remembering the mountain I climbed in Suriname, struggling in the fierce heat with a double-sprained ankle. Somehow I knew that getting to the top of that mountain and listening to the howler monkey's song booming through the forest canopy was the rewriting of my own legend for myself. It meant I, too, could explore the edges of the possible, could peek at the mysteries of life, swim with the piranhas. Prevail. Triumph. Maybe even fall in love.

During that trip I had heard the legends of the Akijos, a tribe who had developed the high technology of their day but chose to use it to dominate other tribes and kill them for complex cannibalistic rites. Up in this cold northern forest, I'm remembering the Bush Negroes, stolen from Africa and made slaves on the rich sugar plantations of the South American Guyanas. They told themselves in song how to escape, and did, and set up their own Africa on a piece of the South American continent. And eventually slavery was abolished.

The Malagasy and the ilmenite mine that is ever closer to being built on their southern shore. They believe that if they keep walking they will always find another tree. But they could also be shown that mahampy will grow if you plant it.

Those men who sailed Darwin's ship around the world for all those years, flogged and chained because they had become drunk on Christmas Day when they might rather have been home with their wives and children. Labour standards like that don't pass muster these days. And what of Hugh Latimer and Nicholas Ridley, prelates in the Church of England and kingly advisers who were burned at the stake on a public street in Oxford during the reign of Mary I because they were not Roman Catholic?

And then there is Oxford University, only in the last few decades opening the doors of the prestigious male colleges to women. Sylvia Earle, who was told she didn't deserve university funds because she was a woman and who became a celebrated marine biologist anyway, without missing a step. Rosemarie Kuptana, who was sent from her home in the High Arctic to a residential school. We know now that was cultural genocide.

We choose to believe. We choose to suspend disbelief. It is not a rational process, but a deeply intuitive one, the rightful heritage of our species' evolution over time. Cannibalism, slavery, the oppression of people because of their skin colour, sex, religion or social class, the certainty that the sun revolves around the earth or that God created all living things in a single moment. All of these have shifted over time for millions of people.

Even businesses can rewrite their legends. British Petroleum, derided by some of its competitors, is among the most famous of a tough crowd on the issue of global climate change. As far back as May 1997, chief executive John Browne said his massive oil-and-gas company was going to take climate change very seriously indeed, unlike most of its major world competitors. 'We have moved – as the psychologists would say – beyond denial,' he said.

He told a crowd at Stanford University on 19 May 1997 that it would be unwise and potentially dangerous to ignore the concern over climate change: 'The time to consider the policy dimensions of climate change is not when the link between greenhouse gases and climate change is conclusively proven . . . but when the possibility cannot be discounted and is taken seriously by the society of which we are part.'

Browne is clear about why BP is listening to the

science. His company needs to make sure its business can be profitable well into the future. This is enlightened corporate self-interest. But underpinning his words is a curious nod to Darwin's biological theories. He acknowledges that the world is not permanent and that life forms are connected with each other. 'Real sustainability is about simultaneously being profitable and responding to the reality and the concerns of the world in which you operate. We're not separate from the world. It's our world as well.'

I wander around in the forest near the cabin, saying goodbye to it for now. My parents are getting old, and they're not sure they can keep the place. This may be my last visit. I hear a rustle. A few metres away, a brood of startled spruce grouse. It's a mother and her four young, their heads bobbing on long necks. They always make me laugh because their bellies are rounded and their legs spindly. They're known as fool-hens around here because they are just so stupid. They eye me for a while, decide I'm benign and carry on their way. Will we wait until all of this is in fragments like so many other ecosystems and then make a fuss about saving the remnants? Will we ever learn?

Several months after I return to work, I get my answer. A group of conservationists, including Monte Hummel, president of the Canadian arm of the World Wildlife Fund, some indigenous Canadians and four mainstream forestry and oil and gas companies have declared that they have reached a private deal and are determined to preserve not just isolated parks, but the whole boreal forest. Half will be worked gently and sustainably by industry. Half will be completely off-limits to commercial development.

It amounts to roughly 530 million hectares, or approximately 53 per cent of the total land area of Canada. It is the biggest conservation commitment ever made anywhere in the world. Far bigger even than what's happened in Suriname. It is stunning.

The part that moves me the most is when I talk with Bill Hunter, president of Alberta-Pacific Forest Industries Inc., one of the people who put together the deal. His father was a park warden, and he grew up in the boreal forest. He lived in aboriginal communities. The beauty of the boreal is plain to him, as it is to me. His company, one of the biggest producers of kraft pulp paper in the world, is large enough that its decision on the boreal ought to make a difference to the forest's future.

He's scared, he confesses to me just before the deal is made public. A few days earlier, he had sat around a table with some of the other business leaders, including the oil and gas producer Suncor Energy Inc., conservationists and aboriginal people. They said to each other: what on earth are we doing? Why are we taking this risk?

They know that they are stepping outside the rules

of the game. Businesses are not supposed to be transparent about their strategy or their prospects. Conservationists are not supposed to be in bed with business. And, after everything that has happened between the world's aboriginal and non-aboriginal people, why should the aboriginals trust in this deal? Hunter sighs. He just believes in this. It's a new way of doing things, and it's right. If not these leaders, then who? And if not now, when?

The deal is not law yet, and not all of the big forestry, mining, and oil and gas companies have agreed to it. Some will probably fight it every inch of the way. It may falter. But it's a huge step. A symbol. And to me it is a sign that the rewriting of the fable of inexhaustibility is already happening.

ACKNOWLEDGEMENTS

This book would never have been born without the extraordinary energies of Anna Porter, my Canadian publisher. You were the first to see that my journeys might be made into a book and encouraged me steadily for two years to plunge in. Writing books, as I was to discover, is quite different from writing newspaper features. Best of all, though, you encouraged me simply to keep going and to follow my own instincts. I think we ended up with something different from what either of us originally imagined, and it was magical to be part of that.

Many, many thanks to Meg Taylor, the superb editor at Key Porter who has helped shape this book and who showed up at all the right times bearing wisdom and pumpkin ice cream. I know it was a labour of love for you, and I'm deeply grateful. Thanks, too, to Janie Yoon of Key Porter. I'll never forget the day you phoned me and declared that the Suriname chapter was 'kick ass!'

Many thanks to Susanna Wadeson of Transworld, whose passion for the planet is infectious. You are simply marvellous to work with. And it was a great joy to enter the extraordinary sphere of Tim Smit of the Eden Project. Your energy and determination add up to a force of nature in themselves. Thanks so much for helping me shape the introduction. And many thanks as well to Mike Petty of Eden, a complexly lovely man who introduced me to the mysteries of Cornwall.

I owe a huge debt to Norman Myers. Not only did you supervise my paper at Green College, Oxford, (which later became the Darwin chapter in this book and part of the introduction) but you also generously introduced me to some of the key ideas that influenced this book. Among them: the work by Ornstein and Erhlich and by C. S. Lewis. Far beyond that, though, you convinced me early on that writing this book mattered.

My profound thanks as well to those who run *The Globe and Mail*, Canada's national newspaper. Most of the chapters in the book are based on articles I wrote for the newspaper while I was its earth sciences reporter. More important than that, though, is the fact that the paper fosters curiosity, passion, rigour and independent thought. Like all great newspapers, it tolerates dissent and respects imagination. I am particularly grateful to Jerry Johnson, *The Globe*'s focus editor, for your great friendship and support over the fourteen years I worked at the paper. You showed me how to find the culturally unknowable and begin to know it. And your office has always been the seedbed of my imagination.

Deep thanks, as well, to Phillip Crawley, publisher of *The Globe and Mail*, who allowed me to be away from the paper to attend Oxford University in 2002 and then take a leave of absence again in 2003 to write this book. You have been unstinting in your praise over the three international awards the paper has won for our groundbreaking work in earth sciences. Eddie Greenspon, *The Globe*'s editor-in-chief, was the one who allowed me to hop on an aeroplane to Madagascar – my first big journey – and write about the sixth extinction. It was an act of faith. That trip changed my life, and I will always be grateful to you for that, Eddie. The bigger gift you have given me, though, is the infectious ferocity of your belief that writing makes a difference, that it matters.

As well, I must thank Richard Addis, who was editor-in-chief of *The Globe* during much of the time I was earth sciences reporter and who understood why it is so important to democratize the growing body of scientific findings on the state of the earth. I still remember the evening I made you supper – salmon and butternut squash and goat cheese – and persuaded you to let me write the series of articles that form the backbone of this book and that eventually won the paper an international award in Berlin for environmental reporting.

Many thanks both to the Reuters Foundation Fellowship Programme, which sent me to Green College, Oxford, and to the World Conservation Union (IUCN) for the generous sharing of information and the depth of your expertise.

I want to thank the biologists, intellectuals and prophets who allowed me to tag along with them on their journeys and ask unceasing questions. The list is long and eminent: Norman Myers, Francisco Dallmeier, Alfonso Alonso, Philip Currie, Rosemarie Kuptana and her mother Sarah Kuptana, Russ Mittermeier, Bragi Arnason, Sylvia Earle. I carry your stories with me still and will savour them until I am old.

Apart from the scientists named in the chapters, though, are all the others who, in effect, gave me private tutorials over many years, painstakingly explaining to me the science of conservation. Chief among these is my father, George Mitchell, one of North America's earliest field biologists and an expert on the pronghorn antelope. I think you fed us ecology at the dinner table, Dad, and I think that's why it made so much sense to me later on. As well, though, I am intensely grateful to Paul Paquet, Steve Herrero and David Schindler.

Writing this book was a challenge in many ways, but the biggest was financial. It was a leap of faith for me to take unpaid time off work to write, and many people helped me over the breach. Glen Davis, you are a god-send and an extraordinarily generous man. And to my parents, George and Constance Mitchell, I can only offer heartfelt thanks. As well, there have been some anonymous donors and lenders whose money came at just the right time.

On a personal level, I want to thank some of the friends who have unfailingly believed in this project – and in me – over the years of research. Jacques Gérin, my fierce first reader who questioned every single grand statement: I could not have written this book without you. John Asfour: you are a marvellous poet but also an unerring editor; you are responsible for the very structure of this book. Teresa Posyniak: your belief in me has quite simply made my life better. Thanks, too, to Lisa Bowen of Conservation International who just kept making things happen.

I want to thank my mother, who listened to every word of every draft of every chapter, long distance, and has kept me going literally day by day during separation, divorce, Oxford, travel and writing. She frequently hopped on planes to take care of my children when I travelled. I guess I know who I get the wanderlust from. To James Patterson, my passionate healer and lover and visionary: you have the power of grace and forgiveness.

And finally, to my children: Calista and Nicholas. This book is for you.

SELECTED REFERENCES

Alvarez, Walter. *T. Rex and the Crater of Doom*. New York: Vintage Books, 1997.

Beer, Gillian. *Darwin's Plots: Evolutionary Narrative in Darwin, George Eliot and Nineteenth-Century Fiction*. London: Routledge, 1983.

Bratchell, Dennis F. *The Impact of Darwinism: Texts and Commentary Illustrating Nineteenth Century Religious, Scientific and Literary Attitudes*. London: Avebury Publishing, 1981.

Brooks, Geraldine. *Nine Parts of Desire: The Hidden World of Islamic Women*. New York: Anchor Books, 1996.

Brown, Peter G. *Ethics, Economics and International Relations: Transparent Sovereignty in the Commonwealth of Life*. Edinburgh: Edinburgh University Press, 2000.

Cain, A. J. 'The True Meaning of Darwinian Evolution'. In Grafen, Alan (ed.). *Evolution and Its Influence*. Oxford: Clarendon Press, 1989.

Carson, Rachel. *Silent Spring*. Boston: Houghton Mifflin, 1962.

Conkin, Paul K. *When All the Gods Trembled: Darwinism, Scopes and the American Intellectuals*. New York: Rowman & Littlefield, 1998.

Constant, Pierre. *The Galapagos Islands: A Natural History Guide*. 5th edn. New York: Norton, 2002.

Darwin, Charles. *The Descent of Man*. 2nd edn. London: Merrill and Baker, 1874.

——. *Diary of the Voyage of H.M.S. Beagle*. Ed. Richard Darwin Keynes. Cambridge: Cambridge University Press, 1988.

——. *On the Origin of Species by Means of Natural Selection, or the Preservation of Favoured Races in the Struggle for Life*. Ed. J. W. Burrow. London: Penguin, 1985.

Davis, Wade. *The Serpent and the Rainbow: A Harvard*

Scientist's Astonishing Journey into the Secret Societies of Haitian Voodoo, Zombies, and Magic. New York: Touchstone, 1985.

Dawkins, Richard. *The Blind Watchmaker: Why the Evidence of Evolution Reveals a Universe Without Design*. London: Penguin Books, 1991.

Dennett, Daniel C. *Darwin's Dangerous Idea: Evolution and the Meanings of Life*. New York: Simon & Schuster, 1995.

Desmond, Adrian, and James Moore. *Darwin: The Life of a Tormented Evolutionist*. New York: Warner Books, 1991.

Earle, Sylvia A. *Sea Change: A Message of the Oceans*. New York: Putnam, 1995.

Eldredge, Niles. *The Pattern of Evolution*. New York: W. H. Freeman and Co., 2000.

Ellegard, Alvar. *Darwin and the General Reader: The Reception of Darwin's Theory of Evolution in the British Periodical Press, 1859–1872*. Gotebord: Goteborgs Universitets Arsskrift, 1958.

Ehrlich, Paul R. 'Human Natures, Nature Conservation, and Environmental Ethics'. *Bioscience* 52 (2002).

Feiler, Bruce. *Walking the Bible: A Journey by Land through the Five Books of Moses*. New York: Perennial Books/HarperCollins, 2001.

Gelbspan, Ross. 'The Heat Is On: The Warming of the World's Climate Sparks a Blaze of Denial'. *Harper's Magazine*, December 1995.

Gould, Stephen Jay. *The Individual in Darwin's World*. London: Weidenfeld and Nicolson, 1995.

——. *The Panda's Thumb: More Reflections in Natural History*. New York: Norton, 1980.

Grady, Wayne. *The Bone Museum: Travels in the Lost Worlds of Dinosaurs and Birds*. Toronto: Viking Penguin, 2000.

Hardin, G. 'The Tragedy of the Commons'. *Science* 162:1243–48.

Hawken, Paul, Amory Lovins and L. Hunter Lovins. *Natural Capitalism: Creating the Next Industrial Revolution*. Boston: Little, Brown and Co., 1999.

Hilton-Taylor, Craig (compiler). *2000 IUCN Red List of Threatened Species*. Cambridge: IUCN, 2000.

Huxley, Julian. *A Book That Shook the World: Anniversary Essays on Charles Darwin's Origin of Species*. Pittsburgh: University of Pittsburgh Press, 1958.

Huxley, Thomas Henry. 'Obituary of Charles Darwin'. *Proceedings of the Royal Society* 44 (1888).

Irvine, William. *Apes, Angels and Victorians: The Story of*

Darwin, Huxley and Evolution. Lanham, MD: University Press of America, 1955.

Jackson, Michael H. *Galapagos: A Natural History*. Revised and expanded edn. Calgary: University of Calgary Press, 2001.

Jolly, Alison. *Lucy's Legacy: Sex and Intelligence in Human Evolution*. Cambridge, MA: Harvard University Press, 1999.

Keith, Sir Arthur. *Concerning Man's Origin, Being the Presidential Address Given at the Meeting of the British Association Held in Leeds on August 31, 1927, and Recent Essays on Darwinian Subjects*. London: Watts & Co., 1927.

Kerven, Rosalind. *The Mythical Quest: In Search of Adventure, Romance and Enlightenment*. Introduction by Penelope Lively. London: Pomegranate Artbooks, 1996.

Kinnon, Collette M. *The Impact of Modern Scientific Ideas on Society*. London: D. Reidel, 1981.

Kuhn, Thomas S. *The Copernican Revolution: Planetary Astronomy in the Development of Western Thought*. Cambridge, MA: Harvard University Press, 1957.

Leakey, Richard, and Roger Lewin. *Origins Reconsidered:*

In Search of What Makes Us Human. New York: Anchor Books, 1992.

——. *The Sixth Extinction: Patterns of Life and the Future of Humankind*. New York: Doubleday, 1995.

Leggett, Jeremy. *The Carbon War: Global Warming and the End of the Oil Era*. London: Routledge, 2001.

Lewis, C. S. *Selected Literary Essays: Bluspels and Flalanspheres: A Semantic Nightmare*. Cambridge: Cambridge University Press, 1969.

Marshall, Peter. *Nature's Web: An Exploration of Ecological Thinking*. New York: Simon & Schuster, 1992.

McKibben, Bill. *The End of Nature*. London: Viking, 1990.

Merlen, Godfrey. *Restoring the Tortoise Dynasty: The Decline and Recovery of the Galapagos Giant Tortoise*. Quito: Charles Darwin Foundation, 1999.

Milman, Henry Hart. *The History of Latin Christianity: Including That of the Popes to the Pontificate of Nicolas V*. London: J. Murray, 1883.

Mittermeier, Russell A., Norman Myers, Cristina G. Mittermeier, Gustavo A. B. Da Fonseca and Jennifer Kent. 'Biodiversity Hotspots for Conservation Priorities'. *Nature* 403 (2000).

Mittermeier, Russell A., Norman Myers, Patricio Robles and Cristina G. Mittermeier. *Hotspots: Earth's Biologically Richest and Most Endangered Terrestrial Ecoregions*. Mexico City: CEMEX, S.A., 1999.

Moore, Berrien III, et al. 'Amsterdam Declaration on Global Change'. From *Challenges of a Changing Earth*. Amsterdam: Global Change Open Science Conference, 2001.

Moore, James. *The Darwin Legend: Are Reports of His Deathbed Conversion True?* London: Hodder & Stoughton, 1995.

Moorehead, Alan. *Darwin and the Beagle*. London: Hamish Hamilton, 1969.

Myers, Norman. *The Primary Source: Tropical Forests and Our Future*. New York: Norton, 1992.

——. *The Sinking Ark: A New Look at the Problem of Disappearing Species*. New York: Pergamon Press, 1979.

Myers, Ransom A., and Boris Worm. 'Rapid Worldwide Depletion of Predatory Fish Communities'. *Nature* 423 (2003).

Novacek, Michael J. (ed.). *The Biodiversity of Crisis: Losing What Counts*. New York: The New Press, 2001.

Ornstein, Robert, and Paul Ehrlich. *New World, New Mind: Moving Toward Conscious Evolution*. New York: Doubleday, 1989.

Pope John Paul II. Message on Evolution to the Pontifical Academy of Sciences. 23 October 1996.

Pritchard, Peter C. H. *The Galapagos Tortoises: Nomenclatural and Survival Status*. Lunenburg, MA: The Chelonian Research Foundation, 1996.

Quammen, David. *Monster of God: The Man-Eating Predator in the Jungles of History and the Mind*. New York: Norton, 2003.

——. 'Planet of Weeds: Tallying the Losses of Earth's Animals and Plants'. *Harper's Magazine*, October 1998, pp. 57–69.

——. *The Song of the Dodo: Island Biogeography in an Age of Extinctions*. New York: Touchstone, 1996.

Raby, Peter. *Alfred Russel Wallace: A Life*. Princeton: Princeton University Press, 2001.

Romanes, George John. *Darwin and After Darwin: An Exposition of the Darwinian Theory and a Discussion of Post-Darwinian Questions*. London: Longmans Green & Co., 1893.

Royal Society and National Academy of Sciences. 'On Population Growth and Sustainability'. February 1992.

Ruse, Michael. *The Darwinian Revolution: Science Red in Tooth and Claw*. Chicago: University of Chicago Press, 1999.

Sobel, Dava. *Galileo's Daughter: A Historical Memoir of Science, Faith and Love*. New York: Penguin Books, 2000.

Stegner, Wallace. *Wolf Willow: A History, a Story, and a Memory of the Last Plains Frontier*. New York: Penguin Books, 1990.

Tuchman, Barbara W. *A Distant Mirror: The Calamitous 14th Century*. New York: Ballantine Books, 1979.

——. *The March of Folly: From Troy to Vietnam*. New York: Ballantine Books, 1985.

Union of Concerned Scientists. *Economists' Statement on Climate Change*. Cambridge, 13 February 1997.

——. *World Scientists' Warning to Humanity*. Cambridge, 18 November 1992.

Vorzimmer, Peter J. *Charles Darwin: The Years of Controversy, The Origin of Species and Its Critics, 1859–1882*. Philadelphia: Temple University Press, 1970.

Weiner, Jonathan. *The Beak of the Finch: A Story of Evolution in Our Time*. New York: Vintage Books, 1994.

Wilson, Edward O. *The Diversity of Life*. New York: Norton, 1992.

——. *The Future of Life*. New York: Alfred A. Knopf, 2002.

INDEX

Aboikoni, Songo, 193–4, 197
aboriginal peoples *see* indigenous peoples
Aeneid, 26, 68, 202–3
Akijos people, 183–4, 190, 202, 268–9
Alberta dinosaur bone beds, 99–109
Alberta-Pacific Forest Industries, 272–3
Alonso, Alfonso, 68, 71–2, 88–93, 277
Altamirano, Marco, 229, 253
Alvarez, Walter, 100
amber, 114
American deserts, 137–8
Amman (Jordan), 122, 132, 134–5, 139, 140
Anglican church, 42, 160, 269
apes, 45, 128, 267
aquifers, 130–40
  Azraq Oasis, 130, 132, 134–5
  Ogallala-High Plains, 137–8
Arabian Desert, 122–40, 215
Arabian oryx, 123–7, 128, 138, 140
Arctic, 147–70, 219
Arctic Ocean, 105, 148, 169, 237
Arnason, Bragi, 213–17, 223, 277
artificial ecosystems, 85
Ashford, Graham, 168
Asidonhopo, 192–3, 195–6

Attenborough, Sir David, 69
Atwood, Margaret, *Oryx and Crake*, 26
Azafady (aid organization), 78–9
Azraq Oasis, 127, 129–40, 215

Baden Powell, Rev. Prof., 45
Bali Barat National Park, 198
Bali mynah, 198
Ballard, Geoffrey, 216
Banks Island, 15, 147–70
  climate change, 147–53, 161–5
  ecosystem, 150–3, 161–5
  hunting, 152, 154, 156, 162, 166–7
  IISD research project, 150–1, 168
  Inuvialuit people, 147–70
  permafrost, 149, 151–3, 169
  Sachs Harbour, 148–50, 154–7, 168–9, 221
  sea ice, 148–50, 161–2, 170
Beagle, voyage of, 28–9, 31–7, 59–62, 94–5, 269
bears
  black, 264
  polar, 152, 155–6, 162, 167
Berenty forest reserve, 90, 92–4
biomes (Eden Project), 22–4
Blue Lagoon spa, 223–6
Bodleian Library, 30, 48
Bodley, Thomas, 48
boreal forest, 263–73
Braman, Dennis, 111–13
Brazil, 29, 61, 174, 201, 205
British Petroleum, 270–1
Browne, John, 270–1
Buckland, William, 115
Bush, George W., 54–5
Bush Negroes, 172, 192–7, 204, 269
Button, Jemmy, 59–60, 164

Canada, 263–73
  Alberta bone beds, 99–109
  see also Banks Island
Central Suriname Nature Reserve, 189–92, 194, 197–8, 200–1, 204–5
Charles Darwin Research Station, 247, 258–60
clear-cutting, 265–6
climate change
  and arctic, 147–53, 164, 213
  and Galapagos, 243
  cretaceous period, 111
  global, 51–4, 266, 270–1
Conservation International, 91, 173–4, 190–2, 194–5, 199–200, 229–30, 240, 248, 259, 279
cretaceous period
  asteroid strike, 99–103, 108–13
  climate change, 111
  extinctions, 75, 108–13, 119

forest, 105–7, 113
K/T boundary, 100–1, 106, 108–11, 119–20
natural selection and, 113
Crofts, Roger, 136
Currie, Philip, 101–11, 117, 119–20, 277
Cuvier, Georges, 115–16

Dallmeier, Francisco, 68, 71–2, 74–5, 83–5, 88–93, 202, 277
Darwin, Charles, 16–19
Beagle voyage, 28–9, 31–7, 59–62, 94–5, 269
Descent of Man, The 45, 267
and dinosaurs, 116–17
and Galapagos, 33–7, 107, 227–9, 231–3, 240–2, 244–7, 261
and Madagascar, 95
and natural selection, 16–19, 36–7, 39–47, 113, 201, 260, 266–7
On the Origin of Species, 17, 28, 39–46, 116, 228
and religion, 28, 32–3, 35–6, 53, 55, 57, 206
on slavery, 29
and Yamana (Fuegian) people, 59–62, 164, 180
Darwin, Robert, 31
Darwin Island (Galapagos), 253, 255, 256, 258
Darwin's finches, 34, 36, 244
Dead Sea, 10–13, 140–5
Deffeyes, Kenneth, 216
deforestation, 50–52
see also forests
Descent of Man, The (Darwin), 45, 267
deserts
Arabian, 122–40, 215
Gobi, 115
Madagascar, 79
north American, 137–8
polar, 149, 153
developing world, 52, 178
dinosaurs
Alberta bone beds, 99–109
archaeopteryx, 116
discovery, 114–16
evolution, 110–13
extinction, 47, 67, 99–103, 111, 138
fossils, 41, 99–109, 114–16, 120
diseases, spread of, 60, 62, 77, 97, 118, 172, 178–9
Dublin Review, 42

Earle, Sylvia, 230–1, 237, 239, 249, 251, 253, 269, 277
ecosystems
artificial, 85
collapse, 50, 102, 108, 111–13, 118–9
endangered, 68–70, 81–2, 124, 126–8, 150, 163–4

ecosystems (*cont.*)
impermanence, 53, 138, 214
islands, 34, 81–5, 93–4, 150–3, 161–5, 228, 231, 235, 242–5
marine, 239, 248, 259
ecotourism, 81, 189
Ecuador, 229, 248
Eden Project, 21–4
Ehrlich, Paul, 19–20, 201–2
Einstein, Albert, 57
Elizabeth I, Queen of England, 49
Ellegard, Alvar, 21, 42–3
endangered species, 176, 177, 199–200, 234–5
Galapagos, 228, 231, 234–5, 242–5
Jordan, 126–7, 131, 134
Madagascar, 68–71, 73–4, 80–2, 88
Suriname, 176–7
energy sources, 52, 66, 144, 211–13, 216–7, 224–5, 243, 272
environmental hotspots, 37, 65–6, 91–2
Esau, Peter, 158, 160, 165, 170
Espinoza, Fernando, 229–30, 253
Evatraha, 76–80, 87, 91, 95, 96
evolution
and natural selection, 16–19, 36–7, 39–47, 113, 201, 260, 266–7
denial, 18, 43–6
extinctions, mass, 50, 67, 68, 75, 95, 102–3
fifth, 102, 108, 117–20
and fixity, 32–3
rates of, 75, 127
sixth, 37, 67, 68, 74–5, 95, 212, 277
Exxon Mobil, 54

Fernandina (Narborough) Island (Galapagos), 242, 247, 250–1
ferns, 113
fishing, 72–4, 76, 166, 170, 239, 248, 255, 258
FitzRoy, Captain Robert, 59–60, 62
fixity of species, 18, 32–3, 53, 115, 117, 214
Fleischer, Ari, 55
flightless cormorants, 234–5, 252
Floreana (Charles) Island (Galapagos), 242, 245, 246–7
Ford, Harrison, 200
forests
boreal, 263–73
Cretaceous, 105–7, 113
gallery, 90, 92–4
littoral, 73, 80
rainforest, 61–2, 73, 171–82, 186, 189–80, 201, 204–5
Fort-Dauphin, 71–2, 76, 82, 90
fossil fuels, 54, 211–17,

265–6, 270–3
fossil water, 134–5
fossils
dinosaurs, 41, 99–109, 114–16, 120
plants, 111, 113, 114
French Guiana, 172, 194, 204
Friends of the Earth, 69, 81

Galapagos Islands, 15–16, 33–7, 107, 227–61
birds, 34, 36, 234–5, 244
Charles Darwin Research Station, 247, 258–60
Darwin and, 33–7, 62, 107, 227–9, 231–3, 240–2, 244–7, 261
climate change, 243
ecosystem, 34, 228, 231, 234–5, 242–5
Galapagos National Park, 229, 233
humans and, 228, 233, 234, 242, 245
invasive species, 234
mangroves, 235–6
marine iguanas, 34, 231–2, 243, 251, 252
sea lions, 235–6, 250–1, 256
tortoises, 34–5, 228, 231, 244–51, 258–60
trees, 235–6, 241, 243
Galapagos National Park, 229, 233
Galapagos penguins, 235–6
Galapagos Tortoises, The (Pritchard), 247
Galilei, Galileo, 43, 57
gallery forests, 90, 92–4
gas *see* fossil fuels
global warming *see* climate change
Globe and Mail, The, 11, 67, 276
Gobi Desert, 115
Gould, John, 36
greenhouse gas emissions, 51, 53–6, 101, 161–2, 164–5, 191, 270
griffins, 114–5
Guyana, 204

Haogak, Edith, 157, 160
Haogak, Phillip, 159–60, 221
harpy eagles, 62, 172, 174–7, 189
Histories (Tacitus), 141
History of Latin Christianity (Milman), 56, 95
Horner, John, 114
Hubbert, M. King, 215–6
Hubbert's Peak (Deffeyes), 216
Hummel, Monte, 271
Hunter, Bill, 272–3
hunting, 50
Banks Island, 152, 154–6, 162, 166–7
Galapagos, 245–6
Madagascar, 79
oryx, 122–4, 127
Suriname, 189
Huxley, Julian, 41

Huxley, Thomas Henry, 40, 41
hydrogen energy, 211–13, 216–17

Iceland, 209–26
  Blue Lagoon spa, 223–6
  cell phone use, 210
  food, 222–3
  geology, 209–10, 223–5
  geothermal energy, 223–6
  hot springs, 218, 225
  hydrogen energy, 211–13, 216–17
  language, 210–11, 219
  Reykjavik, 217–8
  sagas, 210–11, 219–22
  Swartsengi geothermal plant, 223–6
  turf houses, 221–3
ichthyosaurus, 116
iguanadon, 115
ilmenite, 73–4, 76, 80, 269
indigenous peoples, *see* Akijos; Inuvialuit; Trio; Yamana; Yanomamo
industry and environment, 37, 68–70, 81–4, 143–5, 211–18, 226, 270–3
Intergovernmental Panel on Climate Change, 53
International Institute for Sustainable Development (IISD), 150–1, 168
Inuvialuit people, 15, 147–70
Inuvialuktun, 158, 170
iridium, 111, 112
Isabela (Albemarle) Island (Galapagos), 34, 231–4, 242, 252–3
Isch, Edgar, 229
island ecosystems
  Banks Island, 150–3, 161–5
  Galapagos, 34, 228, 231, 235, 242–5
  Madagascar, 81–5, 93–4
Israel, 140, 142–4

jaguars, 62, 172, 185, 190, 192
Jesus College, 48
John Paul II, Pope, 44
Johnson, Daniel, 207–8
Jonsson, Thorsteinn, 223–4
Jordan, 121–45
  Amman, 122, 132, 134–5, 139, 140
  Azraq Oasis, 127, 129–40, 215
  endangered species, 126–7, 131, 134
  extinct species, 129
  reception, 9–13
  Shaumari Reserve, 122–7
Jordan River, 11, 141, 143

K/T (cretaceous/tertiary) boundary, 100–1, 106, 108–11, 119–20
Kaolok, John, 159
Keith, Sir Arthur, 267
Keogak, Donna, 151–3
Keogak, John, 151–4
Kew Gardens, 84

Kuptana, Roger, 155–6
Kuptana, Rosemarie, 150, 159, 164, 168, 269–70, 277
Kuptana, Sarah, 148–9, 157–60, 202, 221, 237, 277
Kuptana, William, 155, 160
Kwamalasemutu, 174, 180–1, 184, 186, 188
Kyoto Protocol, 53–5, 191

Lambert, Daniel, 72–4, 80, 83, 88–9
Leakey, Richard, 67, 74, 202
Lees, Andrew, 69
legends, 25–6, 266–73
  creationism, 28, 42, 45–6, 95, 266–8
  explaining fossils, 113–5, 117
  Icelandic, 210–11, 218–21
  Inuvialuit, 157–60
  Malagasy, 78, 87, 96–7
  sustainability, 56–7, 96–8, 118, 261, 266
  Trio, 180–5
lemurs, 67, 70–1, 79, 81–2, 88–95, 173
Leveillé, Johanne, 72
Lewis, C. S., *Selected Literary Essays . . .*, 25
littoral forests, 73, 80
Lively, Penelope, 96
Lonesome George, 247, 259–60

Madagascar, 14, 65–98
  Berenty forest reserve, 90, 92–4
  desert, 79
  ecotourism, 81
  endangered species, 68–71, 73–4, 80–2, 88
  Evatraha, 76–80, 87, 91, 95, 96
  foreign aid, 66, 82
  Fort–Dauphin, 71–2, 76, 82, 90
  Mandena protected forest, 80–8, 96–7
  Mandena research station, 84, 85–6
  poverty, 66, 71–2, 77–9, 82
  and Rio Tinto, 69, 71–3, 78–85
  slash-and-burn practice (tavy), 70, 79, 91, 96–7
  tree destruction, 70–2, 73, 74, 79–81, 86–90, 92, 96–8
Mandena protected forest, 80–8, 96–7
Mandena research station, 84, 85–6
mangroves, 235–6
marine ecosystems, 239, 248, 259
marine iguanas, 34, 231–2, 243, 251, 252
mass extinctions *see* extinctions
Mast, Roderic, 246–7, 259
McNeely, Jeff, 174

Mediterranean Sea, 144–5
megalosaurus, 41, 115
Milman, H. H., 56, 95
Mitchell, Constance, 278
Mitchell, George, 264, 278
Mittermeier, John, 163
Mittermeier, Russell, 91, 173–207
Moncayo, Juan Carlos, 254
monkeys, 128, 173, 176, 179, 189, 200–1, 205, 268
Moore, Gordon, 200
Museum für Naturkunde, Berlin, 116–7
musk ox, 149–50, 153–4, 156, 162–5, 167–8
Myers, Norman, 27, 37–8, 58–9, 91, 202, 277
Mythical Quest, The (Lively), 96
mythology *see* legends

Nasser, Hazem al, 135
National Academy of Sciences, 51
natural selection
  and Bush Negroes, 192–3
  Cretaceous period, 113
  Darwin and, 16–19, 36–7, 39–47, 113, 201, 260, 266–7
New World, New Mind (Ornstein and Ehrlich), 19–20, 201–2

oceans, endangered, 50, 148–9, 230, 237–9, 248–9, 258
oil *see* fossil fuels
On the Origin of Species (Darwin), 17, 28, 39–46, 116, 228
Ornstein, Robert, 19–20, 201–2
oryx, Arabian, 123–7, 128, 138, 140
Oryx and Crake (Atwood), 26
Owens, Richard, 41, 116
Oxford, 27–8, 41
Oxford University, 11, 49, 57, 115, 251, 266, 269, 276, 277
Oxford University Museum, 115
ozone layer, 58

permafrost, 149, 151–3, 169
Philip, Prince, Duke of Edinburgh, 69
pillars of salt, 142
Pinta (Abingdon) Island (Galapagos), 247, 259
piranhas, 172, 175, 187, 192, 206–8, 256, 268
Plot, Robert, 115
polar deserts, 149, 153
primates, 119
  see also apes; lemurs; monkeys
Pritchard, Peter C. H., 247–8
protoceratops, 114–15

quartz, 111–12

Radcliffe Science Library, 30, 41, 49, 56, 266

rainforest, 61–2, 73, 171–82, 186, 189–80, 201, 204–5
Ramanamanjato, Jean-Baptiste, 90–1
Red Sea, 144–5
red-footed boobies, 234, 253–4
Reuters Foundation Fellowship Programme, 277
Reykjavik, 217–8
Rifkin, Jeremy, 213
Rio Tinto, 69, 71–3, 78–85
Roman Catholic Church, and heresy, 42–3, 48, 269
Romero, Xavier, 233, 255
Royal Society, 51
Royal Tyrrell Museum, 101

Saamaka people, 193–5
Sachs Harbour, 148–50, 154–7, 168–9, 221
Sambo, Clément, 79–80
San Cristobal (Chatham) Island (Galapagos), 241–2
Santa Cruz Island (Galapagos), 247, 258
Santa Fe (Barrington) Island (Galapagos), 247
Santiago (James) Island (Galapagos), 245
Scopes trial, 46, 267
Sea Change (Earle), 239
sea ice, 148–50, 161–2, 170
sea lions, 235–6, 250–1, 256
shark finning, 255
Shaumari Reserve (Jordan), 122–7
shrimp farming, 236
Sigurdardottir, Sigridur, 218–9, 221–2
Sixth Extinction, The (Leakey), 67–8, 74, 102–3, 212, 277
slavery, 171–2, 192, 195–7, 269
  Darwin on, 29
Smit, Tim, 22, 24
Solomon Islands, 172
species, endangered, 176, 177, 199–200, 234–5
  Galapagos, 228, 231, 234–5, 242–5
  Jordan, 126–7, 131, 134
  Madagascar, 68–71, 73–4, 80–2, 88
  Suriname, 176–7
Spix macaw, 205
Sprow, Frank, 54
steppe eagles, 131
Sumerians, 139
Suncor Energy, 272
Suriname (Dutch Guiana), 171–208
  Akijos people, 183–4, 190, 202, 268–9
  Asidonhopo, 192–3, 195–6
  Bush Negroes, 172, 192–7, 204, 269
  Central Suriname Nature Reserve, 189–92, 194, 197–8, 200–1, 204–5
  Kwamalasemutu, 174, 180–1, 184, 186, 188
  rainforest, 61–2, 172–206

Suriname (*cont.*)
Saamaka people, 193–5
slavery, 171–2, 192, 195–7, 269
Trio people, 174–94
Voltzberg Dome, 204–6
Werehpai caves, 180, 183–4, 189
Swartsengi geothermal plant, 223–6
Syrian desert, 123–40

Tacitus, 141
Talbot, Lee, 123–5
titanium dioxide, 73–4
tortoises, 34–5, 228, 231, 244–51, 258–60
Trio people, 174–94

Udenhout, Willem, 190–1
Union of Concerned Scientists, 52–3, 56
United States, and Kyoto Protocol, 53–5, 191
United States National Institutes of Health, 174

Vincelette, Manon, 80–7, 90–2
Voltzberg Dome, 204–6

Wallace, Alfred Russel, 39
water, *see* aquifers; fossil water; oceans; sea ice
Wedgwood, Josiah, 31
Werehpai caves, 180, 183–4, 189
Wolf Island (Galapagos), 253, 255, 258
World Conservation Congress, 122, 173
World Conservation Union (IUCN), 71, 123, 277
World Wildlife Fund, 69–70, 82, 199, 271

Yamana people, 59–62, 164, 180
Yanomamo people, 61–2, 180